WATER RESOURCE ENGINEERING

WATER RESOURCE ENGINEERING

DR. R.N. SAHAY

Oxford Book Company
Jaipur ▪ New Delhi

Published By

Oxford Book Company

Regd. Off. "Bony Residency", Gate No. 2
Opp. Tilak Public School, Vishweswariya Nagar,
Gopalpura Road, Jaipur - 302018, India
Ph: 0141 - 2761280, Telefax: 2761381
email - oxfordbookjpr@gmail.com
website - oxfordbookcompany.com

Delhi Office: 102, 1st Floor, Satyam House,
4327/3, Ansari Road, Darya Ganj,
New Delhi - 110002, India
Ph: 011-45652440

Distributed By

KEYA INTERNATIONAL PRESS LLP
E-mail: keyainternationalpressllp@gmail.com

Published in India, 2025

ISBN : 9789391096496

Typeset by Shivangi Computers, Jaipur

Printed and bound in India by Thomson Press Ltd.,
New Delhi

Contents

Preface

Full efficiency could be achieved by water quality control measures at the individual waste outfall or water supply intake was useful for discussion of the maldistribution of costs resulting from technological external diseconomies, and probably descriptive of a significant number of real cases as well. But engineering-economic research his shown that there are many instances where economies can be realised by collective measures such as joint waste treatment, lowflow augmentation for quality improvement, stream reaeration, groundwater recharge for quality improvement, effluent diversion or redistribution to make better use of natural assimilative capacity, the specialised use of streams or stretches of streams, and combinations of these. Where facilities involve scale economies that cannot be efficiently realised at individual outfalls or intakes, it may be appropriate to have a regional authority with powers more extensive than those assumed in earlier.

At one extreme would be a river basin or sub-basin agency empowered to plan, design, construct, operate, and finance virtually all water quality control measures; at the other would be an agency that simply sees to it through effluent charges or standards-that downstream costs are reasonably well reflected in the upstream decisions of managerially independent units. In this part we emphasize the opportunities which a regional authority might have to improve the efficiency of water quality management by implementing various measures, such as those mentioned above, that cannot be efficiently instituted at individual points of waste generation and discharge. However, we also recognise the importance of the technologic and economic linkages between water quality improvement as an output from water resources systems and other outputs from such systems, such as hydropower, navigation, and water-based recreation. For example, operation of hydro facilities for "peaking" can reduce water quality significantly by causing extreme low flows during "offpeak" periods. Water released from the deeper parts of storage reservoirs which have "stratified" may be devoid of dissolved oxygen and have an effect on water quality downstream from the reservoir similar to the discharge of an organic waste.

The installation of low head dams for slack-water navigation may result in substantial water quality deterioration. The dams reduce the velocity of flow and hold waste loads longer in a given reach of stream; at the same time they reduce the reaeration capacity in the given reach. Conversely, the operation of a storage reservoir to augment flows during summer periods to improve water quality may have adverse effects on water-based recreation on the reservoir, because of the resulting drawdown. These interrelationships suggest the desirability of a regional water resources management agency with responsibility for all aspects of water resources development and use. We return to this point where we discuss more specifically the question of appropriate institutional arrangements for water quality management.

The reader of the book will find comprehensive knowledge with suitable examples to this subject.

— Dr. R.N. Sahay

Chapter 1

Hydrogeological Environmental Assessment

The safe and reliable long-term disposal of solid waste residues is an important component of integrated waste management. Historically, landfills have been the most common, environmentally and economically acceptable method of disposal of solid waste. Even with the implementation of waste reduction, recycling, and transformation technologies, disposal of solid waste in landfills remains a significant component of an integrated waste management strategy. Environmental impact assessment (EIA) can be defined as the systematic identification and evaluation of the potential impacts of proposed projects, plans, programs, or legislative actions relative to the physical, chemical, biological, cultural, and socio-economic components of the total environment. The prime purpose of the EIA process is to encourage the consideration of the environment in planning and decision-making to ultimately arrive at actions that are more compatible with the environment.

In India the environmental impact assessment exercises in the case of landfill projects are still at a development stage. Environmental legislation pertaining to the submission of an environmental impact assessment report for landfills was enacted two years ago. To date, about 50 sanitary landfill projects have been conceived, designed, and completed in India. But consideration of the environmental parameters in designing and developing these projects has been neglected. According to legislation on municipal solid waste-formulated and enacted by the Union Ministry of Environment and Forest as empowered under the Environment Protection Act of 1986 the submission of an environmental impact assessment prior to the designing and development of any landfill facility in the country has been made mandatory.

Any municipal authority or designated agency engaged in the management of municipal solid waste anywhere in the country must perform an environmental impact assessment of the proposed site for a sanitary landfill operation. Landfill projects must comply with the standards for air, water (ground and surface), pollution, and other environmental norms.

The development of a sanitary landfill facility proposed by the local municipal corporation of the city of Jammu is justifiable as ensuring the establishment of a sound solid waste management system. Jammu has experienced massive urbanisation in recent years and the present system of solid waste management has failed to accommodate the quantity of solid waste that is presently being generated in the city. A preliminary environmental impact assessment study for the proposed project has the following objectives:

1. To compile an environmental inventory of the project area. This includes the following details:
   a. Hydrogeological settings
   b. Geographical and meteorological details
   c. Land use pattern of the study area
   d. Ground water quality of the study area
   e. Demographic details
2. To predict the possible impact on groundwater quality
3. To prepare an environmental management plan of the project site

Environmental Inventory of the Project Area

The study area of Jammu district, the winter capital of Jammu and Kashmir state, falls between 32°35'N to 32°55'N and 74°45'E to 75°00'E, covering an area of 870 square kilometres.

Physiography, Relief and Drainage

The study area is comprised of flat land and small hills. The main township is adjacent to lower Shivaliks, which are comprised of lower Himalayan Mountains. The topography is mostly undulating hills. The altitude in the area varies considerably from one location to another, the average height being 500 metres. The main drainage systems in the area are the Chenab, Tawi, Basantar and Ujh rivers. The Chenab is perennial and snow fed. The rest of the rivers are seasonal rivulets known as Khads that traverse the area.

Climate and Rainfall

The climate of the area is subtropical characterised by three distinct seasons: winter, summer, and Monsoon. May and June are the hottest months while December and January are the coldest. The annual average ambient temperature varies from 2°C to 40°C.

Geology

The geology of the area consists of Shivaliks region systems and is mainly composed of sandstone and red transported quartzite. The lower accessible area, including the foothill plains, consists of alluvial deposits left by the seasonal rivulets, which carry away rainy water during Monsoon season. The parent material is mainly composed of alluvium and co-alluvium on the foothill plains. Accordingly, the study area was divided into recent, subrecent, and Shivaliks groups.

Soils

The soils of Jammu show complete heterogeneity. The soils of the foothill and adjoining area are composed of boulders and gravel with ferruginous clay. These types of soils are spread over in the study area and are generally loams, but poor in clay content. Soils on the foothills and V-shaped small valleys have been found to be deep with a medium to heavy texture.

Land Use Pattern

In 1995 around 35% of the region is under cultivation, 37% is forested, 29% is watershed, 0.25% is urban, and 2% of the area is snow cover.

Infiltration Characteristics

The soil infiltration rates vary under different conditions, land uses, and soil types in different hydro-climatic environments. Some recently conducted studies show that initial infiltration rates vary from 0.03 to 2.4 cm/hr for bare ground, 1.2 to 3.0 for agricultural land, 0.3 to 6.3 for grassland, and 0.6 to 1.2 for forestland respectively.

Demographic Details

The pressure on the available resources exerted by the population has an affect on the area. More than 1.2 million people live in Jammu, which comprises about 76% of the total area population; the rest of the population lives in villages. The population density is about 280 persons per square kilometre since the area is very easily accessible and is served by all modern means of transportation while the environmental conditions are conducive to settlement.

Existing Groundwater Quality

The average groundwater level in the Jammu region is at a depth of 18.8 metres below ground level. The average monitoring results of groundwater quality studies show that the general pH of the water varies from 6.5 to 8.5 showing its suitability for drinking purpose. The electrical conductivity values have been shown to exceed the limit of 750-2250 micromhos/ cm. Total dissolved solids have been found to be of 15800 mg/l. Calcium has been found to be within the acceptable limits, which is 250 ppm. The measured nitrate concentration exceeds the limit of 45 mg/l. The reason for these values can be attributed to agriculture runoffs, which result in transportation of nutrients from agricultural fields to water bodies. The fluoride was also found to exist in acceptable limits, which are 1.5 ppm.

Assessment of Hydrological Environment

Assessment of Aquifer Vulnerability using the DRASTIC Method

Figure : Hydrology, environmental protection agency]

The prepared landfill project is supposed to cause some micro-scale impacts, which would be of significant nature on the soil or groundwater within the project area. These significant groundwater impacts are evaluated here using an aquifer vulnerability mapping technique called DRASTIC. DRASTIC is a methodology for identifying vulnerability to groundwater pollution. It uses seven parameters, which are a combination of geologic, hydrologic, geomorphologic, and meteorological factors, to relate an aquifer to the sources of its water and the constituents within that water. The parameters (Depth of water, annual Recharge, Aquifer media, Soil media, Topography, vadose zone Impact, and hydraulic Conductivity) are weighted according to their relative importance in determining the ability of a pollutant to reach an aquifer. The parameters are used to produce DRASTIC index numbers from which maps can be constructed.

The methodology was developed around a set of basic assumptions concerning a generic contaminant. They are:

1. Material introduced at the land surface as a soluble solid or liquid travels to the aquifer with recharge waters derived from precipitation.
2. The mobility of the contaminant is assumed to be equal to that of the groundwater.
3. Attenuation processes are assumed to go on in the soil, vadose zone and aquifer. DRASTIC was not specifically designed to deal with pollutants introduced in the shallow or deep subsurface, such as leaking underground storage tanks or deep mine wastes.

Ratings for each parameter are assigned depending on its contamination potential. Ratings vary from 1 to 10, with higher values describing greater pollution potential. These parameter ratings are used for all evaluations, however, each parameter is also weighted to account for differences between general land use and agricultural use. Weights range from 1 to 5, with higher weights representing greater pollution potential. This technique is based upon the seven parameters, or factors, of the DRASTIC rating scheme coupled with a relative importance weight resulting in a rating for each factor. The factors are:

D = Depth of groundwater

R = Recharge rate (net)

A = Aquifer media

S = Soil media

T = Topography (slope)

I = Impact of vadose zone

C = Transmissivity of aquifer (Conductivity)

Determination of the DRASTIC index number (pollution potential) for a given area involves multiplying each factor rating by its weight and adding together the resulting values. Higher sum values represent a greater potential for pollution or a greater vulnerability of the aquifer to contamination. For a particular area being evaluated, each factor is rated in a scale from 1 to 10 indicating the relative pollution potential of that factor for that area. Once each factor has been assigned a rating it is weighted. Weight values, from 1 to 5, express the relative importance of the factors with respect to each other. Finally the total impact factor score, the DRASTIC index number, can be calculated:

$$\text{Pollution potential} = D_rD_w + R_rR_w + A_rA_w + S_rS_w + T_rT_w + I_rI_w + C_rC_w$$

where

r = Rating for area being evaluated (1-10)

w = Importance weight for the factor (1-5)

Factor ratings are derived from data on each factor while importance weights are found in generic DRASTIC tables that list weights for factors having greater applicability (Aller, et al., 1987). The values of the factors for this study area were calculated from data from the southwestern Himalayan regional centre of the Indian Institute of Hydrology, Roorkee, India.

The data of the various parameters of the study area are as follows:

Depth of groundwater = 32.80 feet

Rate of net recharging = 9 in/hr

Aquifer media = sandstone with pebbles

Soil media = sandy loam

Topography = 5%

Impact of vadose zone = sand and gravel

Transmissivity of aquifer = 1729 pd/ft

This information was calculated into specific values for each factor:

Depth of groundwater = 5

Net recharge rate = 8

Aquifer media = 7.5

Soil media = 6

Topography = 9

Impact of vadose zone = 7.5

Transmissivity of aquifer = 8

These rating values are multiplied by their importance weight then added together to arrive at the DRASTIC index number:

Pollution potential

= DrDw + RrRw + ArAw + SrSw +TrTw +IrIw +CrCw

= 5 x 5 + 8 x 4 + 7.5 x 3 + 6 x 2 + 9 x 1 + 7.5 x 5 + 8 x 3

DRASTIC index number = 162

The DRASTIC index number (pollution potential) is high, thus showing the vulnerability of the aquifer to contamination from the proposed landfill project.

Environmental Management Plan for the Project

The environmental management plan for the landfill facility may primarily address the groundwater contamination and leachate management due to the terrestrial nature of environmental degradation caused by the project. The prevention of groundwater contamination would require the use of geosynthetic clay liners or membranes for the landfill. The contamination probability index calculated by the DRASTIC model was included in the remedial plan for the project. The high saturation of groundwater and the presence of alluvial soil media are the prime reasons that the project poses a contamination threat to the groundwater. Utilising the liner technology of geosynthetic clay materials can mitigate this threat. It was decided to put in a drainage system in order to provide swift and smooth removal of water from rainstorms. The geosynthetic clay liner is provided in sandwiched layers of cement and other stabilisation material. The groundwater monitoring will be done on a permanent basis in order to report any contamination caused by excessive leaching of heavy metals or other chemicals. Besides that leachate and landfill gas monitoring system, management of leachate and minimisation of soil erosion may include some aspects of environmental management of the landfill project.

The present environmental impact assessment of the landfill has been carried out to prepare a database of the impacts that could result from the project. In the present study, the environmental impacts that may occur are analysed with particular emphasis on groundwater contamination. This study has been able to identify the critical deleterious environmental degradation that could occur from the project. The database prepared for the environmental inventory of the project may be useful for other studies in the area regarding the landfill management plan and its implementation during the later stages of the project.

Water Pricing

The claim is often made that more effective use of pricing and various water charges can remedy or completely solve problems of water scarcity, shortage or overuse. A common view is that users pay too little for water and, as a result, they use too much. The public or private agencies that supply water to users could levy higher prices willingly, or the regulatory agencies that manage and oversee water resources could enforce higher rates. Implementing these approaches in developing countries may present special challenges due to the high costs involved or due to inadequate resources to assign, monitor and enforce diverse rights to water use. However, are there reasons that other countries, not so constrained, have not implemented effective water pricing strategies? Is this pricing prescription really so straightforward?

Economists, engineers and others have played a lead role in these policy debates, and economists continue to write volumes about many aspects of the issue. The purpose of this chapter is to review current approaches to effective water pricing and to examine the gap between these academic prescriptions and the current state of water allocation policy and practice. To what extent are more complex or comprehensive systems of water charges likely to fulfill the growing challenges of water scarcity? What steps are needed to bring theory into practice?

The main conclusions reached here are that in many jurisdictions, schemes of administered prices may represent an attractive policy approach to promoting the overall efficiency of water use. The attainment of appropriate or efficient usage levels is synonymous with encouraging conservation and reuse, with investment and innovation in new technologies and practices and with achieving expected levels of water quality and security of supply. It is also consistent with capturing as fully as possible the many beneficial consumptive uses of water in situations where water is relatively abundant or inexpensive. Unlike the allocations that might be reached in some private market transactions for water use, an advantage of administered pricing is that it offers the ability to include, as part of the prices charged, society's best estimates of social costs related to downstream or future users, instream flow uses, and so on. Unlike private market outcomes, administered pricing also offers the ability to make the water pricing program revenue-neutral. A carefully designed rebate program can leave the average water user's annual household or business income unchanged, even while providing strong financial incentives to conserve.

These issues and arguments are explored in the following sections. The discussion starts by examining the concept of water pricing and its rationale, focusing attention on those pricing approaches that promote the efficient use of water resources. The concluding sections highlight pricing issues that require further attention. Some brief illustrations of the state of water pricing in Canada and elsewhere show that despite initial steps toward policy reforms, considerable action is still required to harness the potential gains of effective water pricing.

The Meaning of Water Pricing

Various systems of water rates, water fees and user charges are employed around the world as a means of influencing the processes by which water is provided and used, and as a means of cost recovery or revenue generation. In some places, these water prices are the principal form of water allocation. In other places, these prices are combined with various forms of licenses, permits, quotas, restrictions and other practices and customs that dictate how much water is used, where and at what cost. Where pricing is used, there may be a range of apparent price levels since, in practice, water systems have a number of stages in the supply chain. There might be prices assigned for bulk water withdrawals from a surface water or groundwater source that differ from other prices further along the supply chain. Specific prices might be assigned once the water has been successively transported, stored, treated and distributed for final use by residents, industries, irrigators, public works and so on, and these prices may vary by time, place or purpose of use.

Economists have occupied themselves for centuries exploring the role and behaviour of prices in influencing decisions about the production and use of all manner of goods and services. Long before economists became involved, people had experience with barter and trade in town and village markets everywhere. It was readily apparent that markets could function with varying degrees of effectiveness for a wide range of transactions. At their best, markets and the price signals they generate are capable of coordinating the independent decisions of vast numbers of producers and traders and of allocating scarce supplies of goods and services across diverse consumer groups. It is also the case that markets work more effectively at allocating some types of goods and services than others. Historically, markets have not been the principal means of allocating the use of fresh water in most parts of the world.

Since, in most countries, an individual's ability to use groundwater or surface water is not decided by the interaction of numerous independent buyers and sellers, there is a potentially important distinction to be made between market prices and administered prices. Market prices are those that arise from the changing balance of supply and demand among competing traders, whereas administered prices are those that are set or imposed by some public agency in the absence of an interactive market process. The two types of price determination occupy opposing ends of a continuum, where the determining difference is the degree to which information about buyers' and sellers' valuations drive the price-setting process.

To illustrate this distinction, consider a regional water authority that acts as a single supplier and that sets prices based only on notions of its own cost of service. This pricing behaviour would place it at the administered pricing end of the spectrum, since users' valuations are not reflected in the prices charged. If the water authority's policies and strategies were to change, and the water authority chose to set prices reflective of buyers' willingness to pay, then the resulting market prices might be indistinguishable from those a profit-oriented monopolist also would have charged. Indeed, a central notion for the improvement of water pricing practices around the world is to shift or reorient price-setting objectives and/or processes at the local level. Especially where notions of market efficiency and the best use of scarce resources are paramount, the challenge is to make administered prices more closely resemble the market prices that a well-functioning market might set.

When prices are determined by auction or due to the competitive supply to the market by numerous sellers, an anticipated shortage is usually resolved spontaneously through a price increase. Alternatively, under the administered price approach, a conscious effort would have to be made to adjust prices in times of scarcity. If that effort were insufficient, then other measures such as quotas and water restrictions would be needed if all customers were to be assured some supply. In the face of an anticipated shortage, both pricing approaches do better if there is timely and accurate market information and forecasting. In the case of markets, any adjustment of prices in response to new forecasts is the implicit responsibility of all buyers and sellers collectively. Under the administered approach, the selling agency is the only one that can raise or lower prices. Under both pricing approaches, there remains a risk of shortage. Some of the cost of this risk and of any resulting shortage will be borne by both buyers and sellers.

Discussion about the design and reform of water pricing policy, then, is almost always about the purpose and relative effectiveness of various schemes of administered pricing as a means of allocating usage of water or raising revenue. There may or may not also be a policy option to establish some form of water market. With active water markets, there would be no specific role for pricing policies, since then the specific price levels are not a policy choice but the result of market interactions. Where there is a choice between new or reformed water markets and reformed water pricing, a clear understanding of reformed water pricing would help inform such a decision.

The Rationale for Water Pricing

Setting a price for water use can potentially achieve a number of diverse purposes, but an acknowledged limitation is that each price can generally only achieve one of these purposes and so priorities need to be established.

Typically, the three primary and competing purposes to be prioritised are promoting economic efficiency, generating revenue and advancing economic equity or fairness. Where a water pricing method includes two price elements such as a flat monthly charge plus a volumetric charge based on metred use then it may be possible to find specific combinations of the monthly charge and of the unit charge that can achieve two of these purposes simultaneously. A decision to focus on one prioritised purpose means that other policy instruments are likely to be needed to address the secondary purposes, or else they might go unfulfilled. A decision to not focus on or priorities a single objective could be consistent with a compromise pricing scheme that tries to satisfy all of the objectives somewhat but that is unlikely to achieve any of them fully.

Economic efficiency refers to a water allocation outcome in which nobody could be made better off such as through some other water allocation or water price accompanied by compensating transfers without making someone else worse off. Importantly, this efficiency goal the so-called Potential Pareto Efficiency comprises a number of desirable features or subgoals, including notions of water conservation; water quality improvement; encouraging innovation, investment and risk-taking; and ensuring security of water supply. That is, pricing that seeks to achieve economic efficiency seeks to frame the water allocation challenge quite broadly and to include all of these dimensions of water use that provide value or cost to society. For example, an administered water price that is set too low might encourage too much current consumption and not provide adequate incentives to preserve water for use in later periods or downstream. This would not be an efficient outcome if a higher price could save water and if the gains from subsequent use exceed the near-term sacrifices made to achieve them.

The competing goal of revenue generation may take the form of encouraging a public utility to operate on a not-for-profit basis or of allowing a privately owned utility to earn a fair rate of return on its investment. In other cases, revenue generation from water resources may contribute significant revenues to the public treasury and act as a "tax-by-another-name." Too little revenue generation may result in water use decisions that feature relatively low investment in technology, infrastructure and equipment. Too much revenue generation may result in high prices with too little beneficial use of the water that is available.

The potential mismatch between revenue raising and economic efficiency objectives in setting water prices has generated two main responses among economists. One approach favours choosing economic efficiency as the primary goal and using other public policies, including other taxes or subsidies to meet revenue targets, as needed. The other approach proposes multipart pricing structures more complex systems of water pricing that might be able to achieve both goals simultaneously. Authors in both groups agree that, in some jurisdictions, the historical pursuit of revenue generation objectives as a higher policy priority than efficiency objectives constitutes a barrier to solving water allocation problems that occur there.

The third competing goal of water pricing is economic fairness. Most of the attention here is on ensuring affordable access to some threshold level of water by all users, especially those with low incomes. The other strand of this discussion also seeks to promote fairness by placing high charges or surcharges on consumers of large volumes, under the presumption that such users are relatively wealthy and should pay more. Here there are also two main responses taken in the water pricing literature. One view is that, in both developed and developing country examples, the status quo is unwarranted. Specifically, low-priced or unpriced water is providing a greater benefit to relatively high-income households than to low-income households. Under such pricing schemes, wealthy users often gain access to more abundant, reliable supplies at a relatively low cost. Poorer users are often under-serviced and may go without water, use less water or pay more overall. In developing countries, the cost to the poor includes travel time to distant sources and high unit prices for purchased water not reliably available from the low-priced public supply. In developed countries, a low unit charge is sometimes accomplished by also charging a (correspondingly higher) monthly flat rate, and the overall effect disadvantages the poor.

The second response to this focus on fairness is to note that, especially in developed countries, the annual cost of threshold amounts of water use even under higher pricing that promotes economic efficiency would be a relatively small part of a low-income household's budget. For instance, these water payments would often be less than what is paid for access to energy, transportation, housing or entertainment. Where ability to pay is an issue, it is best to address it through income support measures and safety net programs, not through the de-linking of water costs from water use.

How should this controversy over the primary intended purpose of water pricing be resolved? Especially in jurisdictions where there are significant current or anticipated problems of water overuse, scarcity or quality, it seems likely that the value of water pricing as an instrument of public policy will yield its highest social return if it is directed toward the pursuit of economic efficiency. In this situation, goals for revenue generation and fairness are more likely to be met using available policy instruments other than water prices. Of course, the primacy of the economic efficiency goal and the lower relative priority and advantage of the competing goals will vary by jurisdiction, and could be compared using standard economic methods for evaluating policy choices. Sometimes, there will be multiple possible schemes to price water efficiently, such as when considering whether to supplement a per unit charge for water with a fixed monthly fee. In these cases, revenues and fairness need not be ignored. In general, decisionmakers considering a pricing policy should be careful not to jeopardize efficiency by saddling their program with revenue-raising targets as well.

In those cases where efficiency is the main goal of water pricing, there may also be other criteria that influence how and whether pricing reforms are implemented. These include policy flexibility, administrative feasibility, political acceptability and overall effectiveness. These criteria can play a role in the design and implementation of new reforms and in the ongoing assessment of established pricing systems. Historically, in some jurisdictions, water rates have rarely been adjusted, yet fluctuating water supplies and demands may recommend a flexible price regime accompanied by timely information dissemination about current prices to individual users. The flexible regime would make water pricing more closely resemble the way that some households and businesses contract for electricity and other energy in fluctuating markets. Seasonal pricing or more frequent rate adjustments would allow for users to adjust their own water uses according to its current scarcity value at any point in time. Relatively complicated pricing systems, such as those that vary by volume, time and point of use, require greater administrative capacity, including investments in monitoring or smart metering. In some jurisdictions, these services will be too costly or are not available.

Types of Pricing Promote Efficient Water Use

There seems to be agreement among economists about the principles for rate setting, even if these practices are not yet widely implemented. Without having to address concerns about revenue generation or fairness, the role of prices is to serve as an incentive for users and suppliers to reach quantities of use that are efficient. That quantity of use is one for which the value to the user of the last unit of water used equals the costs of supplying it incurred by that user and everyone else the so-called social marginal cost of supply. This cost will vary according to the source of the water and the time and place it is supplied, among other factors.

There is one principal advantage of administered water pricing compared to allocating water entirely through markets, auctions and tradable entitlement schemes. With administered prices, there is the option to employ a more comprehensive and relevant measure of supply costs than those reflected in most markets. For example, suppose users withdraw surface water from a river system. One should look beyond their costs of pumping or diversion to define a marginal water cost that also includes costs incurred by downstream users who will access less water. Suppose users withdraw groundwater from aquifers with storage capacity. In defining these social costs, one should ask not only about opportunities foregone by other current users, but also about costs to future users who will access less water. Two non-market valuation methodologies provide an excellent starting point for estimating such values. So-called stated preference and revealed preference methods can generate these estimates for incorporation into appropriate definitions of social marginal cost and administered water prices.

Attempts to reach the optimal levels of water use will require estimates of how these costs vary over space and time. In almost all parts of the world, there are relatively predictable patterns of water flow and use that depend upon climate and hydrology, as well as upon established cycles and patterns of human activity. Consider the situation in which there are one or more wetter months or seasons, and one or more relatively water-scarce months or seasons. To achieve efficient outcomes in seasons of water abundance, one should expect to see a lower price of water that reflects the lower marginal costs of providing it. Since this water is relatively abundant, the set of relevant costs might include only the costs of pumping, distribution and treatment, if any, but relatively little cost in terms of displacing or crowding out the demands of others (e.g., downstream or instream users). Conversely, in dry seasons, the relevant costs include the same wet-season elements plus all the associated costs of storage and of other system expansion that allow these dry season users to be provisioned with the expected degree of reliability and water security. The result will be a pattern of administered water prices that could vary significantly from month to month, for example. Under such prices, users would face strong incentives to conserve and to innovate in water saving, including optimal investments in methods and techniques that conserve or reuse dry season water. Similarly, users should know that when water is abundant, they will be allowed full access at low or zero prices, provided those prices cover the relatively lower costs of their current use in those months.

These pricing principles describe an approach to reaching efficient usage levels of water in times of both shortage and abundance. Even though these are administered prices, they are capable of providing important signals across the economy about preferred patterns of water use and about alternative means of supplying or provisioning water. For example, Grafton and Kompas examine the way that Sydney, Australia, has coped with periodic droughts and water shortages without using water pricing effectively. When there is a shortage, Sydney authorities impose various schemes of water use restrictions. These restrictions impose costs indiscriminately on broad groups of residents and water users. Historically, there has not been much public interest in investing in permanent or standby water capacity to address periodic droughts and shortages. According to Grafton and Kompas, Sydney should impose prices sufficiently high to ration available supply during these drought episodes. The authors' estimates show that, at these prices, it would be immediately apparent that investments in alternative standby sources, such as desalination, would be warranted, even if the facility is expected to be used infrequently.

Notice that in no part of this discussion of pricing to promote efficient water use are systems of increasing or decreasing block rates involved, under which neighbouring users pay different rates according to their levels of monthly use. Indeed, if the cost of supply is the same, then efficiency is hampered if high block prices are arbitrarily curbing the water use of one household while allowing the household next door to consume at lower price levels. Whatever their other merits, such as for revenue raising or the pursuit of fairness, there seems to be agreement that these rate design features do not promote efficient water usage levels.

Higher water prices that are introduced to support efficiency goals might, in practice, generate significant revenues. This revenue collection is a side effect that is itself not necessary for higher or lower prices to guide users to efficient allocation decisions. If revenue generation is unpopular, authorities can exercise their option to implement revenue-neutral water pricing. Under such a scheme, authorities commit to returning all revenues to the water consuming population, provided that the rebate amounts are not directly related to actual water use. For example, the rebates could be used to lower property taxes or to eliminate the monthly flat rate portion of water and other utility bills.

Although there seems to be wide agreement about how these principles should apply to the pricing of water at both the wholesale and retail (residential) levels, some aspects of water pricing may not be well understood. Three issues that deserve brief mention here are differences between surface water and groundwater, water and sanitation linkages and the concept of user cost. The foregoing prescription for water pricing is to adopt a broad definition of marginal cost incurred by society in using incremental units of water. These marginal values should differ according to whether the source of the water is groundwater, surface water or some combination of the two. It is important not to overlook these cost differences, such as by using some average price or cost. Increasingly, water resource managers have come to understand that groundwater aquifers can play an important storage and conveyance role, and that groundwater may have extra value as a supply of last resort if surface water should become unavailable. Accurate price signals that reflect these differences can help to coordinate optimal systems of conjunctive water use. That is, the optimal response to a higher relative price for the groundwater could include using groundwater mainly in times of surface water scarcity and ensuring full groundwater recharge in times of surface water abundance. Further, all customers could gain from having serviced access to water from both sources, allowing customers to benefit from whichever is cheaper.

Especially at the residential level, there is a strong linkage between household water usage and household sewer and sanitation throughput. Increasingly, water usage can be metred and billed at low cost, but sanitation throughput cannot. There are some differences across households according to exterior (lawn and garden) water usage that affect the relation between the two service flows, but these can be estimated fairly directly. An issue for many jurisdictions is that sewage treatment, particularly at the tertiary stage when phosphates are removed, can be quite costly. Its cost may not be well understood or appreciated by water-using households and thus not reflected in households' consumption behaviour or choice of appliances, for example. Authorities can adjust their use of metred water pricing principles to reflect the complementary provision of sewage treatment services at the household or business level. The effect will be to increase the overall efficiency of the two services combined, likely reducing water usage levels in the process.

Water pricing practices and principles may not have paid enough attention historically to the cost of raw water, whether provided from groundwater or surface water sources. Raw water refers to untreated water drawn directly from natural water sources. Many water pricing regimes that focus on only cost recovery do not see raw water as a cost and do not charge for it. If these agencies adopt pricing principles based on the long-run marginal cost of water, they might still overlook it. Economists argue that, just like the other costs incurred along the way, one must include any cost incurred from using this "free" water now.

Economists refer to these costs as marginal user costs. In the case of storable groundwater, the cost of using some of the water now is that water might become more costly to pump from greater depth or to treat for poorer quality in the future. In the case of surface water, using more of the water now might take away valuable reserve or standby capacity, necessitating the expense of developing other storages or supplies sooner than would otherwise be the case. Either way, these are costs to estimate and to include as part of an effective pricing regime. Given the accumulated writings and debates on the merits of alternative water pricing schemes, to what extent does one see evidence of their adoption already in practice?

Water Pricing in Canada and Elsewhere

Despite Canada's abundant freshwater, dais water is not always available when and where it is needed. Approximately 60 percent of the country's freshwater drains to the north, away from the 85 percent of the population living near the country's southern border. Due to economic growth, the demand for water in Canada has been increasing rapidly, posing a major challenge for water use and management. In some parts of Canada, water resources have become increasingly scarce and a constraint to the further expansion of those economic activities that heavily rely on water.

There have also been concerns about water quality stemming from contamination by municipal, agricultural and industrial activities. Poor water quality poses a threat to human health and increases the costs of using it productively Water resource managers now face concerns about the impacts that global climate change could have on water resources in some regions.

In Canada, the federal and provincial levels of government share the jurisdiction for water management, with some delegation by provinces to regional and municipal governments. Water prices differ widely across provinces and across municipalities within each province.

At the provincial level, systems of permits regulate water use. Typically, there are prices charged for these permits to withdraw water, but additional permits may not be obtainable in many cases. Water allocations are more likely to be restricted by the quantity provisions of these permits than by the permit fees that are charged. Seven of the thirteen provinces and territories charge some type of one-time or recurring fee for permits to withdraw water. In the remaining provinces, there is no charge at all for them. These fees charged vary from $0.01 to $143 per 1,000 cubic metres of permitted annual water withdrawals.

The concerns about water resources in Canada are exemplified by the recent initiatives that the provinces of Alberta and Ontario have undertaken to improve water management. In Alberta, the govermnent's Water for Life strategy aims to provide safe drinking water, to maintain and protect aquatic ecosystems and to manage water in ways that support sustainable economic development. Part of this policy framework includes a commitment to examine more effective uses of economic instruments for water management, potentially including water pricing.

In Ontario, concerns about water use have motivated the introduction of a new water charge for industrial and commercial water users. Commercial and industrial water users who withdraw more than 50 cubic metres per day from groundwater or surface water sources or from a municipal system will be required to pay a charge of $.00371 per cubic metre of water starting in 2009. The money collected from the charge will cover a portion of the provincial costs of water management activities.

In addition to the fees levied directly by provinces, there is a remarkable diversity of water rates and water rate structures charged by regional and municipal water authorities, such as for water use by residents and businesses.

The main categories of rate structures are flat rates, constant rates, decreasing block rates and increasing block rates. It is estimated that 37 percent of households in Canada face a flat rate charging system (i.e., they pay a fixed monthly fee that is independent of the amount of water they use) and 39 percent of households face constant water rates (i.e., the unit price does not vary with the amount consumed). A further 13 percent of households face water rates where the unit price decreases with the amount consumed, and 10 percent face water rates where the unit price increases with the amount of water consumed. Note that for more than one-third of Canadian households with municipal water service, payments for water service are not in any way connected to the amount of water consumed.

Current systems of water pricing have come under criticism for diverse reasons that relate to the range of roles that water pricing can potentially play: efficiency, revenue generation and fairness. With respect to efficiency, there are concerns about the public's inability to obtain new water permits in regions where water is scarce and complaints about the general lack of effective water conservation measures or incentives. With respect to revenue generation, much concern is expressed when, in many cases, the revenues generated by water prices are insufficient to meet all the operating and capital needs of water suppliers.

One study estimated that only about 50 percent of these costs were being met. There are concerns about the fairness of some historical allocative provisions, such as those that allow holders of senior appropriative water rights to consume vast quantities of water without any payment, even under severe drought conditions. There are public expectations that some new water pricing system should solve all of these problems, yet this is simply not feasible.

The approach suggested here is to focus pricing incentives on issues of water efficiency, and to address concerns about utility finances and about fairness through other available policy actions. Fortunately, price increases motivated by efficiency concerns may also help with revenue targets and fairness concerns, but are unlikely to resolve them completely.

Canada is not alone in expressing discontent about a wide range of water pricing structures and practices. As Young and McColl report, Australia is also struggling to find the appropriate water pricing scheme, even in the face of the nation's pervasive water scarcity and drought. Commenting on a recent survey by Australia's National Water Commission, Young and McColl describe the water pricing practices of fifty-seven urban water suppliers. Of these, twenty-five charge a flat service fee plus an inclining block rate (volumetric charge), twenty-four charge a flat service fee plus a constant volumetric charge, four use an inclining block rate without the flat service charge, one uses a declining block rate, and only three apply the flat service charge independent of the water volume used. As with the examples from Canada, if there is a pattern or principle to these varying pricing practices, it is not apparent.

A recent economic commentary on Australia's water woes concludes that there is a "long-term mismatch between demand and supply that can only be resolved by an increase in prices." In the European Union, water pricing has come to be regarded as an important tool that can be used to improve the management of water resources. The European Union Water Framework Directive (WFD) came into force in 2000 and aims to make the management of European water resources more efficient.

The WFD aims to create new tools for sustainable water use, including measures for managing water demand and patterns of water use. Article 9 of the WFD obliges member states to take into account the principle of recovery of the costs of water services including environmental and resource costs. Article 9 further specifies that members have to ensure by 2010 that water pricing policy is an incentive for efficient water use and thereby contributes to the environmental objectives, and that different water uses make an adequate contribution to the recovery of the cost of water services. This approach to water pricing that the WFD mandates has considerable potential to resolve current inefficiencies in water management and to bring about a water use regime that is sustainable.

Despite the WFD water pricing principles, the actual water pricing practices followed in the member states differ considerably. So far there has been a mixed experience with the integration of economic and environmental objectives into the pricing policies of the member states.

Overall, full recovery of financial costs is only partly achieved, and the environmental and resource costs are rarely considered. Water pricing policy faces its greatest challenges in the agricultural sector, especially in southern European countries, where agriculture is by far the largest consumer of water and where water scarcity problems are the greatest. In the European Union, as elsewhere, pricing is not the sole policy instrument that can or will solve numerous water resource problems, but it would appear its full potential has yet to be realised.

In many jurisdictions, schemes of administered prices are likely to represent an attractive policy approach to promoting the overall efficiency of water use. The attainment of appropriate or efficient usage levels is synonymous with encouraging conservation and reuse, with investment and innovation in new technologies and practices and with achieving expected levels of quality and security. It is also consistent with capturing, as fully as possible, many beneficial consumptive uses in situations where water is relatively abundant or inexpensive.

Unlike the allocations that might be reached in some private market transactions for water use, an advantage of administered pricing is that it offers the ability to include, as part of the prices charged, society's best estimates of social costs related to downstream or future users, instream flow uses, and so on. Unlike private market outcomes, administered pricing also offers the ability to make the water-pricing program revenue-neutral, such as by returning some or all of the fees collected to the water consuming population in any manner not directly related to actual water use.

Chapter 2

Maintaining the Status Quo: Protecting Established Water

For decades, water law in all four Pacific Northwest states has been based on the prior appropriation doctrine. That doctrine has been a fixture in the western United States for over a century, and the fundamental rules of water law based on prior appropriation are well established. These basic rules provide water users with a high degree of certainty and security, creating private property rights to use a resource that is owned by the public, but they also limit water use in some significant ways. For example, the traditional rules restrict where, when, how, and how much water may be used, and specify how water rights may be established and lost. The Northwest's water resources and those who rely on them have come under increasing stress in recent years. Causes of this stress include overappropriation, drought, population growth, the decline of salmon and other fish populations, aquifer depletion, water quality impairment, assertion of tribal water rights, increasing competition for water supplies, public demands for environmental protection, and other factors. These factors have helped create and exacerbate water conflicts, often forcing the states to face difficult issues they had long avoided.

In facing these issues, the states commonly have not applied the traditional rules of western water law. Instead, states often have effectively waived or abandoned these rules in order to preserve existing water-use practices. State deference to existing water uses takes many forms, from silent yet unmistakable failures to enforce long-standing rules, to changes in state statutes for the express purpose of allowing established uses to continue. The accommodation of status quo water uses is now the overarching principle of Northwest water policy. Simply put, the Northwest states protect water users' established practices more faithfully than their legal rights. Where existing water uses are inconsistent with traditional rules of water law, the states have often bent, changed, or ignored those rules in order to preserve these established practices.

This unwritten policy favours those appropriators whose uses would be curtailed by a straightforward application of water law principles and disadvantages others, including other existing water users, prospective new users, and those interested in restoring instream flows, who would benefit if these principles were implemented. Many of the traditional rules of western water law are fading in practical importance as legislatures, courts, and agencies protect existing water uses from the application of these rules.

The states' policy of protecting existing uses seems to be based on considerations of politics, economics, and equity. But this policy may damage other water users, impair instream flows, limit economic opportunities, and ultimately create greater uncertainty about water use and management.

Water Use in the Northwest

People not familiar with the Pacific Northwest, and even some who are, often believe the region is far wetter than it really is. The region's legendary rainfall is not evenly distributed, and the area east of the Cascades is typically quite dry. In Idaho, average annual precipitation in the twenty inches. Even west of the Cascades, where winter precipitation is generally abundant, many streams run short of water in the summer and fall. In the Dungeness and Quilcene basins on Washington's Olympic Peninsula, for example, low streamflows are a chronic problem. Given the demands for water to meet existing and new uses, both instream and out-of-stream, most of the Northwest does not have enough water.

Irrigation is, by far, the dominant out-of-stream water use in the region. Irrigation accounts for over eighty-seven percent of total water withdrawals in the Northwest and nearly ninety-seven percent of the region's water consumption. The next largest water use, public supply, accounts for less than five percent of the withdrawals, while industrial withdrawals are less than three percent. The percentages vary somewhat by state. In Idaho, for example, irrigation accounts for ninety-five percent of water withdrawals and livestock use exceeds public supply and industrial withdrawals combined. In Washington, the most urbanised of the Northwest states, public supply makes up eleven percent of water withdrawals and industrial use more than six percent, but irrigation still exceeds seventy-six percent of total withdrawals. Water withdrawals have sharply depleted stream flows in many Northwest rivers. As in other parts of the West, state efforts to protect instream flows have been too little and too late to keep rivers from drying up. Also, water diversions and low streamflows have seriously harmed fish and other aquatic and riparian organisms. Inadequate instream flows related to water withdrawals are a continuing and significant problem for western rivers and aquatic ecosystems. These water use patterns of the Northwest are not a recent development. They have been firmly established for decades, within the framework of state water laws based on the bedrock principles of the venerable prior appropriation doctrine. The following subsection briefly examines some of these principles.

The Traditional Basic Rules of Western Water Law

Western water law under the prior appropriation doctrine essentially is based on a handful of fundamental and long-standing rules. These few traditional rules are clearly established and simple to state, if not to apply. In general terms, these rules follow.

State Control of Waters

Perhaps the most basic rule of western water law is that the states control the use of water, which state laws declare to be a public resource. States provide for private rights to use water, but those rights may be established only as authorised by the states. Moreover, exercise of such rights is subject to state administration and enforcement.

State Approval of Water Rights

The traditional steps needed to establish a water right are intent, diversion of water from its natural source, and application of water to a beneficial use. Today, however, new appropriations may be made only on the basis of a permit issued by a state water resource agency.

First in Time, First in Right

A water right's priority is based on the date of appropriation, and the oldest rights have the highest priority. Where water is insufficient to satisfy all rights, junior appropriators' withdrawals are curtailed so that users with senior rights may continue to receive their full water supply.

Beneficial Use Without Waste

Water may be appropriated only for a specified "beneficial use," and water rights authorise the use only of enough water to satisfy that beneficial use. In addition, water uses must be reasonably efficient; no one has a legal right to "waste" water.

Appurtenancy

Water rights usually specify a particular place, as well as type, of beneficial use. In most states, water rights are appurtenant to a specific parcel of land, and may not be used elsewhere without a transfer.

Conditional Transferability

The holder of a water right may change its point of diversion, place, or purpose of use, but only with prior state approval. Any such change, however, must not injure the rights of other appropriators, whether junior or senior to the right being transferred.

Use it or Lose it

A water user may lose her water right by failing to exercise it, either by nonuse of water for a period of years (forfeiture) or nonuse coupled with evidence of intent to abandon the right (abandonment)

The foregoing list does not attempt to identify all the rules of western water law, or provide a detailed analysis of individual rules. It does, however, provide a brief summary of the most fundamental and familiar points of the prior appropriation doctrine. In theory, these traditional rules form the basis of water law in the Northwest states. As the following section explains, however, these rules are often sacrificed for the protection of existing water uses.

The Status Quo Policy in the Northwest

The Northwest states remain officially committed to the prior appropriation doctrine. In reality, however, the states often effectively waive the traditional rules where they threaten to curtail established water uses. States have routinely bent, changed, or ignored the traditional rules, through action or inaction, in order to maintain such uses. As a practical matter, water management in the region today is based less on strict prior appropriation principles than on protection of the status quo, that is, preservation of the water use practices, legal or not, that have become established in an area.

"Where is the inconsistency?" the reader may ask. "Isn't prior appropriation itself based on protection of existing uses?" Certainly the traditional rules of western water law protect these uses in many ways, but the rules also establish limits to that protection. For example, users who violate the limits of their water rights by exceeding the specified rate, duty, or season of use, or applying water to unauthorised lands, are not protected by the traditional rules. Nor are users protected who fail to use their water rights for an extended period, who waste water, or who do not observe legal requirements for establishing or maintaining a water right. Junior appropriators are not protected from being shut off in favour of seniors during times of shortage, even if the juniors have not previously been subject to such regulation. And the traditional rules do not prevent water from being used in new places or for new purposes, so long as such changes injure no other appropriators and are approved by the state. Clearly then, the traditional rules of western water law do not provide absolute protection to existing water uses.

Two points must be noted at the outset of this discussion of the states' practice of protecting established water uses. First, there is no inherent conflict between preserving the status quo and applying traditional water laws. To the contrary, maintaining the status quo is fully consistent with prior appropriation rules so long as 1) all users' rights are clearly established, 2) water has been consistently well managed (based on good data) and distributed among those users by priority, and 3) all appropriators are using water in accordance with the terms of their rights and with reasonable efficiency. However, to the extent that these conditions are not fully met, then current practices may deviate from the law. Where water rights are somehow in question, good data are lacking on basin hydrology or actual water use, or there is inadequate water use regulation, then existing uses may differ dramatically from users' legal rights. Depending upon the circumstances, some users will be better off with the status quo, while others will benefit if water is used and managed "by the book."

Second, while the Northwest states usually preserve the status quo at the expense of water law principles, there are exceptions. The results are not entirely consistent largely because state governments are not monolithic. State water-resource agencies, citizen commissions and boards, legislatures, governors, trial courts, and appellate courts all have jurisdiction over water matters. Often, one state government entity has sought to preserve an established water use by changing or not applying the law, while another entity of the same state government has tried to uphold the law. For example, the governors of both Oregon and Washington have vetoed recent legislative efforts to revise state law in favour of status quo water uses. When the Idaho Department of Water Resources refused to curtail groundwater use to protect senior appropriators, the state Supreme Court held that the agency had a mandatory duty to regulate. In Washington, when the Department of Ecology took this same action, the state supreme court held that the agency had no authority to do so. Given these common divisions within a state government, it is somewhat imprecise to speak of "state" efforts to protect established water uses, and defenders of the status quo do not always have the last word.

Having stated these caveats, this Chapter examines the recent actions and practices of Northwest states with respect to established water uses. The record shows that the states commonly pursue an informal, unofficial policy of protecting the water use status quo even when it conflicts with traditional water law principles. This status quo policy, like prior appropriation, has a few fundamental rules. Unlike the basic features of prior appropriation, however, these rules are difficult to locate in any treatise or casebook on western water law. Instead, they must be synthesized from state efforts to protect existing uses. With rare exceptions, states abide by the following basic precepts: 1) enforce the law only when necessary, 2) change the law where needed to protect existing uses, 3) avoid the position of having to curtail established water uses, and 4) prevent instream demands from threatening existing out-of-stream uses. This section discusses these basic tenets of the status quo policy and examines how states have implemented them.

Enforce the Law only When Necessary

In deciding whether to curtail ongoing water uses, the water resource agencies in the Northwest states take a passive approach. If "enforcement" is defined as requiring water users to comply with applicable laws and water right conditions, such as duty or place of use, all four states take enforcement action against water users almost exclusively in response to complaints from other users. Montana's official policy is that a "complaint is always required unless the regional manager can document compelling reasons for recommending enforcement without one." Even when disputes arise between groups of existing users, the agencies can be reluctant to step in.

In Whatcom County, Washington, the state Department of Ecology (Ecology) has taken little action against widespread illegal water use, despite a 1993 survey that found over five hundred users taking water without a valid right. Most of these users were irrigators, and many had been using water illegally for decades, without a valid state permit, decreed right, or precode claim. Despite this clear violation of state law, Ecology took no meaningful enforcement action against the Whatcom County users. Washington Governor Gary Locke vetoed a bill approved by the 1997 Washington legislature that would have granted amnesty to the Whatcom County users. Locke did, however, sign a bill that allowed these users to file water right claims that they had already relinquished under state law, effectively authorising the uses to continue.

It is impossible to ascertain the true extent of noncompliance with water laws and water rights in the Northwest, largely because the state water resource agencies do not actively seek out violators. A perception exists that violations are common, and that enforcement is infrequent or ineffective. For example, the Snake River Salmon Recovery Team stated "[t]he dewatering of salmon breeding and rearing habitat must be eliminated. The monitoring of screen use, screen performance, and the quantity of water diverted must be greatly improved." But most of the information is anecdotal, and there has been no systematic study of water law violations and enforcement.

Recent events in the Wallowa River Basin of northeast Oregon illustrate the problems of ascertaining noncompliance. An anonymous writer claiming to work for the Oregon Department of Fish and Wildlife sent a series of letters to state officials and a Portland television station alleging widespread and serious water right violations in the Wallowa Basin. The letters claimed that data obtained from state sources showed actual diversions often far exceeding legal limits. The Oregon Water Resources Department (OWRD) responded that these allegations were based on mistaken assumptions about the limits of Wallowa Basin water rights, and denied that there was significant illegal use in the area. Department officials acknowledged, however, that they had very limited enforcement staff in the Wallowa Basin since the basin's watermaster position had been discontinued in the 1980s. Given this lack of staff, the Department itself may not know whether all water users are in compliance.

The environmental group WaterWatch approached state officials in 1996 with concerns about excessive water diversions in the Wood

River Basin of southern Oregon, but the Water Resources Department essentially denied that illegal diversions were occurring. WaterWatch then employed a hydrologist who used a flow metre to take measurements of Wood River diversions. The hydrologist reported back to the state that there did appear to be serious illegal water use in that basin. State officials continued to dispute the WaterWatch findings, but acknowledged that "the concerns raised by WaterWatch have some merit. We believe, as does WaterWatch, that the lack of adjustable headgates in some cases, or lack of easily read measuring devices in other cases, can result in excessive diversions" Shortly thereafter, the agency ordered Wood River irrigators to install adequate headgates.

Oregon law actually authorises the Water Resources Department to allow illegal water uses to continue temporarily. Specifically, the Water Resources Director may issue a "limited license," a sort of revocable temporary water right, in connection with an enforcement: order to address an illegal water use. Essentially, the law authorises the director to condone an otherwise illegal water use, provided the director orders the user to comply with the law within a specified period of time.

The Montana Department of Natural Resources and Conservation (DNRC) has an unofficial policy of deferring to informal arrangements (such as water-right pooling or rotation agreements) among local water users on a stream, even when those arrangements are "extralegal." DNRC staff believe that such arrangements are common in Montana. DNRC not only tolerates these "extralegal" arrangements, but actually supports them, primarily because DNRC believes they are usually well tailored to the hydrologic conditions of the watershed, and local water users like them.

None of the Northwest states have taken strong action to implement the requirement of beneficial use without waste. As one commentator recently concluded:

The law in Idaho, Montana, Oregon and Washington requires that water not be wasted but there has been no meaningful enforcement of this requirement in any of the states. In all four states, the failure to enforce appears to be because the states lack 1) information on actual water use, 2) a clear waste definition, and 3) political support or wherewithal for anti-waste enforcement. The consequences of this failure include injury to other legal water right holders, both instream and out-of-stream, as well as harm to the public's rights as the owners of the water resource.

When the Northwest states have made limited efforts to encourage efficient water use, they have been very deferential to current water uses and practices. None of the four state water resource agencies in the Northwest believes it has adequate resources to do its job. When asked by agency headquarters for an estimate of additional staffing needs, one OWRD field manager responded as follows:

You asked for a "Cadillac" and a "Citron" [sic] version of what staffing additions are needed to regulate to rate and duty on all Snake River tributaries. Our needs estimate is more of what you might call a "Chevy" or "Dodge" version, like the earlier Grande Ronde proposal. This "Chevy" version is believed to allow us to regulate to rate and duty in two years.

We think that this is reasonable and justifiable. The minimum needs estimate is just that, a minimum. This staffing level will not get us to rate and duty, but is hoped to be sufficient to let us identify problem areas. Not much more than [the watermaster's] description of our current program, "a one speed bicycle."

Despite increasing pressures on water resources in the 1990s, none of the state legislatures have provided any significant increases in agency enforcement resources. The states are not the only water managers in the Northwest with a poor record of enforcing legal requirements; the U.S. Bureau of Reclamation (Bureau), too, has been passive in ensuring compliance with federal laws and contracts. The Bureau has known for well over a decade of widespread misuse, or unauthorised use, of water from federal reclamation projects. This "water spreading" arose in part because the Bureau had consistently failed to ensure that project water was being delivered in accordance with legal and contractual requirements. In 1994, the Bureau promised that it would act to resolve water spreading, but soon its officials were assuring project irrigators that it had taken the wrong approach. Shortly thereafter, the Bureau quietly abandoned the issue over the protests of environmentalists.

The agencies' failure to enforce is somewhat understandable, in part because any efforts they make to curtail existing water uses may be vitiated by the legislatures or courts. Consider the experience of OWRD in its attempts to regulate certain groundwater users whose pumping affected surface flows in the Umatilla River Two water users filed suit, and the district court for Umatilla County blocked the Department from regulating them in conjunction with surface water. In addition, the 1995 Oregon legislature approved a bill intended to block such regulation, although the governor vetoed it.

Change the Law Where Needed to Protect Existing Uses

Central to the status quo policy is the notion that existing laws are generally less important than existing water uses. As explained in the following paragraphs, the Northwest states in the 1990s enacted numerous revisions to their water laws that help preserve established water uses. In addition, the 1997 Washington legislature approved several bills that would have significantly altered state water law in favour of status quo water uses, although Governor Locke vetoed all or part of ten such bills.

The Idaho Supreme Court's 1994 decision of Musser v. Higginson created something of a crisis in the state. Under prior appropriation law, the Musser case was hardly revolutionary; the court unanimously held that the Idaho Department of Water Resources (IDWR) had a clear legal duty to distribute water in accordance with rights of prior appropriation. Musser threatened a serious upset of the status quo because it called on the state to curtail groundwater use from the Snake plain aquifer in favour of senior surface water appropriators. Consequently, IDWR quickly adopted conjunctive use rules that, while giving a nod toward prior appropriation, authorised (continued) "reasonable use" of both surface and groundwater resources, Thus, after the Idaho Supreme Court forced a reluctant IDWR to curtail junior groundwater users, IDWR established more "reasonable" rules, that is, rules more favourable to junior pumpers than the traditional tenets of prior appropriation.

The Idaho Legislature has also acted repeatedly to legitimize existing, otherwise illegal irrigation uses through the Snake River Basin Adjudication (SRBA). The legislature protected water users who had irrigated an excessive number of acres, or who had changed their place of use without the necessary state approval, Idaho enacted statutes in 1985 and 1989, attempting to grant such illegal uses a presumption of validity, but the meaning of these laws was exceptionally opaque, and the court in In re SRBA declared them unconstitutionally vague in 1994. Within weeks, the Idaho legislature passed new statutes that clarified and strengthened protection for otherwise illegal uses in the SRBA. The legislature declared that the public interest was served by confirming past expansions and transfers of water rights that had been made out of compliance with state law. The Idaho Supreme Court upheld the new statutes as constitutional.

Montana and Washington also changed state statutes to resurrect certain forfeited water rights. Montana law originally prescribed a deadline of June 30, 1983 to file claims of existing rights in the statewide general adjudication. Users filed a mind-boggling 216,000 claims, and the adjudication rumbled forward. Nevertheless, in 1993 the Montana legislature authorised the filing of "late claims," essentially granting amnesty to those who had lost their rights by failing to file ten years earlier. Washington took the same action three times, enacting statutes in 1979, 1985, and 1997 to allow the filing of precode water right claims, even though state law had provided that such claims were conclusively deemed waived and relinquished if not filed by 1974. Both states thus reinstated old water rights that had already been forfeited by operation of law.

The Oregon legislature acted specifically to preserve existing water use by the Grants Pass Irrigation District, and to save the district's embattled Savage Rapids Dam. Savage Rapids is an aging diversion dam on the Rogue River that poses a serious barrier to salmon passage, and also diverts water at a far greater rate than the district needs to irrigate its acreage. Oregon officials found that the district's diversions exceeded its water rights, and refused to issue a new right allowing the practice to continue. After years of conflict, the district struck a deal with the Oregon Water Resources Commission and dam opponents: the district would receive a temporary water right, allowing continued diversions at a somewhat excessive rate, while the district had to begin taking steps to remove the dam and replace it with pumps. The Commission issued a permit on the basis of that agreement in 1994. But the following year, the Oregon legislature approved two bills in an effort to save Savage Rapids Dam. The first would have legislatively issued the district a new water right with no conditions regarding dam removal. Oregon Governor John Kitzhaber vetoed that bill, chiefly on the grounds that the legislature should not issue water rights. The second bill blocked dam removal temporarily while a new task force, established by the bill, studied the matter. Kitzhaber signed that bill into law, effectively derailing the water right/dam removal agreement.

The Oregon legislature first authorised "amnesty" for water users in 1989, establishing a process whereby irrigation districts could legitimize otherwise illegal changes in their water use, subject to certain conditions. By the end of 1995, Oregon had created several more exceptions to the appurtenancy requirement, particularly for irrigation districts. Among other things, Oregon established a streamlined process for the transfer of water rights within irrigation districts, including water rights already subject to forfeiture for five successive years of nonuse. A 1995 statute allowed districts to obtain a certificate of water right for irrigation of lands not covered by the underlying permit. The legislature did, however, preserve the "no injury" test for all these changes.

Avoid the Position of Having to Curtail Uses

Agencies, courts, and legislatures have recognised that they can often preserve the status quo quite effectively by limiting state involvement in water rights controversies. This approach not only allows existing uses to continue, but also saves agency resources and reduces state entanglement in potentially nasty matters. Limiting state involvement may not be good public policy, but it often makes sense in both bureaucratic and political terms by keeping agencies and elected officials from taking unpopular stands against established water uses.

The Washington Supreme Court's 1993 Sinking Creek decision, Rettkowski v. Department of Ecology, severely constrained Ecology's authority to administer water rights. In the Sinking Creek case, Ecology had stepped in (after about twenty years) to regulate junior groundwater irrigators in favour of senior surface water ranchers. A majority of the Washington court, however, held that Ecology had no statutory authority to regulate water use until a general stream adjudication had been conducted in the basin. Two justices entered a persuasive dissent on legal, practical, and policy grounds. Among the points made by the dissent:

The majority correctly points out that its decision will not provide for a "cheap and easy" water adjudication solution. Prohibitively expensive and interminable litigation is what the majority has fashioned as a solution, and to no purpose. The relief sought by neither party was for a general adjudication, and yet that is now the only relief which the majority opines is available. The director of Ecology, upon reading the majority opinion, will surely scratch her head in wonderment that she has the responsibility for issuance of water use permits but no authority to regulate those permits.

In the years since the Sinking Creek decision, the Washington legislature has done nothing to restore Ecology's regulatory authority over water use. Thus, for the majority of Washington river basins not yet adjudicated, the state exercises no significant control over existing water uses. Idaho is currently undergoing its own "expensive and interminable litigation" to adjudicate the water rights of the Snake River Basin, which covers nearly the entire state. A 1996 decision of the SRBA court, however, raised profound questions about whether the adjudication would actually scrutinize current water use in any meaningful way, or simply confirm "paper rights." SRBA-Judge Hurlbutt ruled that Idaho statutes did not allow for "partial forfeiture" of water rights. In other words, forfeiture of a water right for nonuse would be an all or nothing proposition; unless completely forfeited, water rights would not shrink after obtaining their maximum size. The Idaho Supreme Court overturned this decision, holding that Idaho law does provide for partial forfeiture. Had Judge Hurlbutt's ruling stood, it would have resulted in SRBA decrees for water rights considerably in excess of current uses. At the very least, it would have reduced agency and court involvement in major disputes over forfeiture, which would not have become an issue unless a right had been entirely unused for several years.

In Montana's massive statewide adjudication, DNRC has scaled back its efforts to evaluate and oppose water use claims. The agency is less active and aggressive than it used to be, largely due to pressure from the water court itself. Such a policy may in fact speed up the adjudication, which has long been an interest in Montana, but it also protects existing users who generally have the most to lose from a thorough evaluation of their water rights. As one commentator has stated:

Water right adjudications promise secure rights, finality, and better management of water resources. But uncertainty and ineffectual state control favour the status quo ante users, most of whom are also in a position to control the pace of adjudications through litigation and the effectiveness of adjudications through state legislatures. Claimants in control of large blocks of water have only lukewarm enthusiasm for decrees and stridently challenge any decision or order that diminishes their claimed rights. Every set of water users faces the possibility that a final decree and better administration may limit or interfere with their accustomed water use. The State of Oregon, as a policy matter, has chosen not to regulate water use in the

Klamath Basin until the Klamath adjudication is complete. At the same time, Oregon has essentially taken the position that the U.S. Bureau of Reclamation, which owns and manages the Klamath reclamation project, may not deliver water for any purpose but irrigation until the adjudication is complete. According to the state, the Bureau lacks authority to deliver water for fish, tribal needs, or wildlife refuges in the Klamath Basin, because the project's water rights are exclusively for irrigation. Thus, the state claimed that it had the sole authority to allocate water and regulate its use, but declined to exercise that authority. Oregon's stance may be legally suspect, but it is politically and bureaucratically expedient. The state's position is at once pro-existing user and anti-federal government, and while asserting state authority, it actually limits state involvement. Some observers believe that watershed councils and other locally driven, consensus-based efforts have gained popularity in the Northwest largely because they can help guard the status quo. Whatever their potential benefits for resource protection and restoration, local consensus efforts seem likely to protect existing water uses, limit state involvement, and reduce controversy on contentious issues. The trend toward local empowerment runs counter to traditional western water law, however, in that it tends to "localise" a resource that is legally controlled by the state and owned by the public.

Prevent Instream Demands from Threatening Existing Uses

Water users often voice support for instream flows, but user groups typically do not want to see streamflows protected and restored at their expense. Thus, the status quo policy moves states to ensure that instream flow demands do not threaten existing uses, either through some specialised legal approach or under the traditional rules. This aspect of the status quo policy is perhaps best illustrated by the basic character of state instream flow programs in the Northwest. The laws of Idaho, Oregon, and Washington all essentially limit the establishment of minimum protected streamflows to state agencies, while Montana's law is just slightly less restrictive. These arrangements effectively allow politically powerful user groups, with their influence in state government, to maintain considerable control over instream flow protections. These programs also relegate instream rights to second-class status in a system that one commentator has deemed "socialist":

Barriers still stand to the full integration of instream uses into state water allocation systems. Barriers (such as the "second class" treatment of instream flows) are not due to any natural incompatibility between instream rights and the prior appropriation doctrine. Far from it; the doctrine is fully up to the task of accommodating modern needs. All that remains is to eliminate those legislative restrictions which are essentially alien to the doctrine's nature.

The most pervasive and frustrating limitation on instream water rights in most western states is the prohibition against ownership of instream rights by persons other than a designated state entity. This prohibition is a curious twist on the prior appropriation doctrine. It reflects a basic discomfort with the concept of instream rights and an underlying distrust of those entities which may seek to acquire them, particularly environmental groups and the federal government.

Ironically, then, the drafters of such programs have turned away from the market-based principles which underlie the prior appropriation system, and have embraced principles of command-and-control resource allocation socialism, if you will with respect to instream rights.

Conceptual problems aside, the Northwest states' instream flow programs face many practical obstacles. These programs must overcome procedural burdens, inadequate enforcement, minimal funding, and political pressure in order to protect streamflows in any meaningful way. These limitations undoubtedly reflect existing users' discomfort with instream flow rights, even where those rights would be later in time and thus lower in priority. Washington Governor John Locke vetoed portions of a bill passed by the 1997 Washington Legislature that would have largely transferred instream flow protections into local hands. The bill would have established local "planning units," with multiple votes for local governments and water users, but only one vote for the Washington state government. These planning units would have had the authority not only to determine future instream flows, but to "adjust" streamflow protections already established by the state. But for the Locke veto, all of Washington's existing minimum flows could have been eliminated, watershed by watershed, by vote of local governments, water users, and property owners. The vetoed legislation would have shifted to the local level a large measure of the state's traditional control over a public resource, although the legislation would not have allowed local planning units to "adjust" existing out-of-stream water rights. Private transactions to convert existing water rights to instream use, whether permanently or temporarily, could provide some senior water for flow restoration. Each of the Northwest states makes some provision for such transactions, while maintaining significant limitations. The 1995 Montana legislature authorised a time-limited program allowing temporary transfers of water for instream flow use; a cautious step, certainly, but still a step in the direction of expanding options for flow protection. In Oregon, by contrast, bills were introduced in the 1995 and 1997 legislative sessions to prohibit any transfer of water from irrigation to any other use. The U.S. Bureau of Reclamation supported changing Idaho law to allow the transfer of natural flow water rights for instream purposes, but it could not even get the bill introduced in the 1997 Idaho legislature.

Environmentalists sometimes suggest eliminating wasteful water use as a promising method for restoring depleted streamflows. The Washington Supreme Court recognised that possibility in State Department of Ecology v. Grimes, but seemed to foreclose it. In Grimes, the court reviewed a referee's report that seemed to consider environmental factors in evaluating the Grimes' claims in a water rights adjudication. The court noted that, while considering environmental impacts is consistent with certain state obligations under the water laws, "these factors cannot operate to impair existing water rights." In fact, the court specifically rejected the use of a "reasonable efficiency" test for reviewing existing water rights, saying it was inappropriate in the context of an adjudication.

The Idaho Supreme Court similarly rejected the application of the public trust doctrine in the Snake River Basin Adjudication (SRBA). The court confirmed that water rights in Idaho are held subject to the public trust, which can take precedence over vested water rights. But according to the court, the public trust was not an element of a water right, and therefore not an issue to be determined in the SRBA. The Idaho legislature acted quickly to address a perceived threat to established water uses by passing a statute in 1996 that purports to preclude application of the public trust doctrine to water rights.

Effects of the Status Quo Policy

The basic rules of western water law work to the benefit of some groups of people and the detriment of others. Whatever else might be said about them, the traditional rules at least give all interest groups a clear sense of what to expect if the rules were in fact applied. Senior users would have a reliable water supply, but would be limited to the terms and conditions of their water rights. Junior users' diversions would be curtailed, if needed, to satisfy senior rights, but senior rights could not be changed or expanded to the juniors' detriment.

Instream flows would have very late priorities, but they would nonetheless be protected to that extent. They would benefit over time as rules, such as waste and abandonment, were applied against consumptive rights. Prospective new users would be unable to obtain a new water right from a fully appropriated source. However, they would have an opportunity to acquire an existing right and change its use, perhaps transferring the water to a new type of consumptive use or even instream use. But under the status quo policy, the results are often quite different from these expectations.

Winners and Losers Change

When compared to what would happen if the traditional water law rules were applied, if a state preserves the water use status quo, it often helps one class of existing water users to the detriment of another. A legislature, court, or agency may understand that it is helping one interest at another's expense, but in some instances may not realise that it is effectively redistributing water from one group to another.

Many decisions essentially have involved a choice among irrigators: surface water users versus groundwater users, juniors versus seniors. Between these classes, the winners have not been consistent, but the states have consistently preserved the status quo. In other words, the states have chosen to allow water users to continue their established practices, rather than apply the traditional rules of prior appropriation in a manner that would change those practices. The Oregon and Idaho legislatures effectively chose seniors over juniors (who would have benefited from the enforcement of water laws and water rights) when they offered a sort of amnesty to many existing water uses.

On the other hand, the Idaho Department of Water Resources's conjunctive use rules, issued in response to the Musser decision, effectively favoured junior users who pumped groundwater over seniors using surface water. The Washington Supreme Court maintained the status quo, thereby helping junior irrigators at the expense of seniors, when it took a narrow view of the Department of Ecology's authority in the Sinking Creek case. In these latter two instances, both seniors and juniors had long used water with little state involvement. The seniors would have benefited from curtailment of the juniors' pumping in times of shortage, but the Idaho rules and the Sinking Creek decision created obstacles to state regulatory action.

The Sinking Creek decision illustrates how the status quo policy can harm even senior water users who have tried to defend their rights. The Sinking Creek ranchers held senior rights to surface water, and, although they began complaining to Ecology in the mid-1960s that they were being harmed by junior irrigators' groundwater withdrawals, the agency took no action until 1989. When the agency did take action, the junior pumpers sued to block regulation, and the Supreme Court ultimately held that Ecology lacked statutory authority to act. Thus, as the Sinking Creek dissenters noted, the seniors were left with no good options to settle their long-running dispute. The dissenters openly urged the Washington legislature to fill the management void created by the majority opinion, but so far the legislature has not done so.

Protection of established water uses may also harm prospective new users. By refusing to apply laws that would reduce the quantity of water legally held by existing users, states limit the amount of water potentially available for new appropriations. For example, Montana and Washington could have freed up some water for new uses by applying their forfeiture laws against users who had failed to submit timely claims, but the states chose instead to protect the status quo by allowing the filing of late claims. Even though improving efficiency could provide water for new uses, none of the Northwestern states has taken serious action against waste.

States also may disadvantage prospective new uses by restricting transfers of existing water rights. For example, Montana prohibits the transfer of large water rights unless the applicant can prove that several conditions are satisfied. Oregon constrains interbasin water transfers to protect the basin of origin. Idaho law prohibits changing the purpose of use of agricultural rights if the change "would significantly affect the agricultural base of the local area." Bills introduced in the past; two sessions of the Oregon legislature would prohibit any water transfers from irrigation to any other use.

Instream uses probably suffer most under the status quo policy. Like prospective new consumptive uses, instream flows generally lose out when states refuse to require existing uses to comply with established laws and water right conditions. Flow restoration efforts are also undermined by laws and attitudes restricting water right transfers and leases for instream use. For example, Idaho law precludes transfer of consumptive water rights for instream use, even though the state water plan calls for that prohibition to be changed. Moreover, states sometimes disarm instream flow advocates by preventing them from employing certain legal tools. For example, the Idaho legislature effectively shut conservation groups out of the Snake River Basin Adjudication. Similarly, the Washington Supreme Court in Grimes held that consideration of environmental values could not "operate to impair existing water rights."

Even in a crisis atmosphere arising from the prospects of salmon run extinction and increased federal regulation, the Northwestern states seem unlikely to take strong enforcement action against existing users in order to protect streamflows. A recent "crisis" in Oregon arose from the proposed listing of two stocks of coho salmon under the federal Endangered Species Act. State political leaders pushed hard against a listing, urging the federal government to defer to Oregon's own plan, the Oregon Coastal Salmon Restoration Initiative. The state initiative contained several measures relating to water quantity and instream flows, addressing new instream and consumptive water rights, enforcement of existing rights and laws, increased water use efficiency, and other matters. Oregon's plan did not commit the state to taking aggressive measures on existing uses. For example, the initiative stopped short of requiring water users to measure and reports their withdrawals, promising only that the state would consider it. The initiative called for four additional Water Resources Department field employees, but did not commit the Department to taking controversial enforcement actions such as limiting water users to their annual duty or regulating groundwater in conjunction with surface water. The state agreed to convene "collaborative" local groups to develop water use efficiency standards, but was vague as to whether it would establish standards in the absence of local consensus, or how it would use them in any event.

Continued protection of existing user groups will likely prevent meaningful progress toward restoring streamflows in most of the Northwest. The water use status quo means seriously depleted (even nonexistent) flows in many rivers and streams. New or recently established streamflow protections, at best, will merely prevent the situation from getting worse. Water right acquisitions for instream flows and public funding for water conservation offer win-win solutions that seem likely to work far better in theory than in practice, given the massive amounts of money required to restore flows in even one river. Economic and environmental costs will also limit new storage projects and other "structural solutions" to flow restoration. Cooperative local efforts based on consensus, for all their promise and recent popularity, offer little hope of addressing instream flow problems effectively. The conclusion is unpleasant but inescapable: protecting the water use status quo means accepting the continued dewatering of Northwest rivers and streams for the foreseeable future.

In general, existing water users in the Northwest have been remarkably successful at continuing their established practices even where these practices violate venerable principles of western water law, and even though applying that law would often benefit other interests. The reason for existing users' influence in maintaining the status quo is open to debate, but public choice theory offers an interesting insight. One branch of public choice theory suggests that groups representing narrow economic interests are often remarkably effective in securing favourable results from public bodies. Research on interest groups' influence in Congress shows that a group is most likely to succeed when the group 1) is attempting to block, rather than obtain legislation; 2) has goals that are narrow, with little public visibility; 3) is supported by other groups and public officials; and 4) can shift the issue to a favourable forum, such as a friendly congressional committee.

This research suggests some of the reasons why existing users often win; they seek to preserve the status quo (although they sometimes have done so by having legislation passed, rather than blocking it), they have limited goals in low-visibility issues, they typically have strong support from their legislators and other political figures, and they often have favourable venues in legislative water committees and local courts, if not in the state agencies themselves.

Economic Questions Persist

The state and federal governments commonly justify the protection of the water use status quo on economic grounds. Thus, Idaho offered "amnesty" to otherwise vulnerable uses in the Snake River Basin Adjudication to protect "significant investments by water users and tax base for local governments." Similarly, the U.S. Department of the Interior has been reluctant to attack subsidies or even unauthorised water uses, because users have come to depend on them. The rationale appears to be that individual livelihoods and local communities have come to rely on existing water use arrangements and it would be unfair or unduly costly to disrupt them, regardless of the problems they pose.

Certainly there are some people who have come to depend (directly and indirectly) on existing water uses, and their lives would be disrupted if those uses were discontinued. In other contexts, however, this rationale is not enough to justify continued violation of the law, or especially continued expenditure of public funds. The Umatilla Tribes have noted the disparity in treatment between illegal tribal fishing and illegal irrigation: "When Indians fish illegally, we are sent to federal prison. When irrigators kill fish by illegally taking water, they are not punished. Instead, we are told by the United States Government that we must consider the impacts to the irrigation economy of making them stop their illegal activity." Military bases, too, have closed despite their economic importance to local communities. The fairness of protecting existing uses based on economic reliance is arguable, depending on one's point of view.

Even from a purely economic standpoint, the status quo policy seems unsound in many cases. On one hand, existing water uses certainly provide economic benefits to their users, their local communities, and the Northwest states. User groups and their allies emphasize these benefits in defending the status quo. On the other hand, existing uses impose various kinds of costs on society, both by causing harm to valued resources (such as dewatering salmon habitat or degrading water quality) and by preventing water from being used for other purposes (such as hydropower generation, new businesses, or instream flows). A comprehensive and accurate look at the costs and benefits of water uses may show that many current uses are far from optimal.

Critics of the status quo point to a variety of natural resource problems associated with current water uses. Such problems carry economic costs, although these costs are hard to quantify because of difficulties in valuing ecosystem services and other benefits of resources "in place." Clear identification and valuation of such costs and benefits would force the Northwest to reassess the economic worth of many current water uses. The public value of some uses may be far lower than commonly believed. Even economically beneficial uses, for example, irrigation of high-value crops in Idaho, may prove suboptimal at the margin, because society might benefit if some portion of the water used for irrigation went to other purposes instead.

Critics also maintain that, compared with alternative water uses, existing uses often generate little economic return per unit of water. Irrigation of alfalfa and other forage crops is most easily questioned, because these crops have relatively low values and high water demands. The marginal value of irrigation water in the West is commonly estimated to be less than fifty dollars per acre-foot. Meanwhile, municipal water suppliers have been willing to pay upwards of $2000 per acre-foot in recent transactions across the West. In some places, the highest economic use of water may be instream. From a purely economic point of view, perpetuating lower valued water uses is not efficient. As one commentator has stated, "although individual users, whether in cities or in rural areas, are probably efficient from their point of view, at the system level the vast differences in the relative value of water and often neglected instream values suggest that inefficiencies abound." Facilitating transfers from current uses to new uses of higher value, including instream uses, could produce real economic and environmental benefits.

Here again, public choice theory recognises that special interest groups may obtain outcomes that are not economically efficient from society's point of view. As scholars point out, this is not necessarily a bad result, because economic analysis is an imperfect tool for evaluating public policy questions of fairness and other issues must also be considered. But if the status quo policy produces economically inefficient results, decision-makers should openly and clearly explain why protecting existing users is in society's best interests. One interesting aspect of the status quo policy is that it sometimes contradicts that commonly stated maxim: "water flows uphill toward money." While that maxim seems likely to hold true over the long run, the current trend is to preserve existing uses. One commentator has observed that recent attempts to block out-of-stream or basin transfers amount to "quasi-riparianism." Money can sometimes make water move uphill, but existing users often can stop water from moving anywhere. Thus, existing water users may be even more influential than money.

Implications of the Status Quo Policy

Despite the pressures of population growth, environmental restoration, and economic transformation, established water uses have remained solidly protected in the Northwest. In order to perpetuate current uses, state legislatures, courts, and agencies alike have refused to apply, and sometimes have even changed, legal requirements. The record clearly shows that the states have been extremely reluctant to implement the traditional rules of western water law if doing so would mean upsetting the water use status quo. Today, those traditional rules are often honoured in the breach; if they ever were sacred, certainly they are not any longer.

In managing water resources, the Northwest states rarely deviate from the status-quo policy. Like it or not, anyone with a stake in the resource should recognise that the states generally will do what they must in order to protect existing water uses. Legal rules, agency mission statements, and even the interests of other water users usually are not enough to force action that would threaten established uses. Given this reality, stakeholders and decision-makers should approach water-resource issues somewhat differently than they would under the traditional rules. No one, not even the owner of vested senior water rights, should assume that her interests are safe simply because she is entitled to protection under basic water law rules. Those rules routinely have not been applied or have been changed, harming junior appropriators sometimes, seniors sometimes, and instream flows nearly every time.

Anyone with an interest in a water-resource must actively and vigilantly defend it, or risk having that interest impaired for the sake of some other user group. In addition, anyone considering a proposal, strategy, or scenario involving water use should evaluate it on the assumption that the status quo will be protected. For example, in evaluating whether a state management plan that emphasizes water law enforcement is sufficient to avert an Endangered Species Act listing, the federal government should ask: Is this likely to work? Will the state actively enforce the law against existing water users and make it stick? A group of irrigators considering a lawsuit to protect their water supply against other users should ask: Can I expect a court to shut off existing users? If so, will the legislature step in to protect them?

A state agency establishing policy for water use during droughts should ask: If we waive certain rules for water users during drought periods, will we be able to reinstate those rules when the drought is over? Or will the users find a way to continue the practices they started during the drought? Obviously, it would be foolish to assume that the law will never mean anything at all. But one should question any approach that relies on any level or any branch of government to enforce the law against established water uses. Finally, while the states obviously have great concern with protecting existing water uses, they otherwise seem to have little regard for preserving western water law. The elements of traditional water law that remain most strong and vital water rights as property rights, which last forever unless abandoned, with priority based on seniority are the elements that most effectively safeguard established uses.

Most of the other elements are either moribund (such as the prohibition against waste), rarely enforced (forfeiture), or readily changed (appurtenancy), To the extent that the traditional rules remain relevant, it is not because they provide a legal framework of any inherent rightness or power. Instead, they persist largely because water politics in the 1990s still favours existing water users who defend their current practices and continuing entitlements based on economics and equity.

Traditional western water law does not directly recognise political, economic, or equitable concerns. But by managing water based largely on these factors, states tacitly recognise that prior appropriation forms an imperfect and incomplete basis for water policy and management. In a sense, when states go outside the traditional rules of western water law to preserve the status quo, they sacrifice prior appropriation for a kind of "public interest," even where the interest being protected is a private water use that actually violates current laws. In sum, by consistently choosing to protect established water uses rather than applying the familiar rules of prior appropriation, the Northwest states have significantly undermined those rules. The region's water law and policy is now based more on today's notions of politics, economics, and equity than on any set of long-standing, bedrock principles. These notions will eventually change, however, and the water users now benefiting from the status quo policy may someday wish that the states had been more faithful to the traditional rules of western water law.

Chapter 3

Western Instream Flow Issues and a New Water Future

Water is the West's most precious resource. Water management, water law, and water policy in the West are at a crossroads. In 1889, during the Montana Constitutional Convention, John Wesley Powell proposed that the state of Montana adopt watersheds as county boundaries. Powell believed that governance should be based on "watershed commonwealths." His advice, offered over one hundred years ago, fell largely on deaf ears. The complicated and often controversial water allocation decisions currently facing stakeholders in the West offer a once in a lifetime opportunity for Westerners "to create a society to match its scenery." This opportunity is arising in watersheds around the region, from the Klamath River in Oregon and California, to the Lemhi River in Idaho, to the Blackfoot River in Montana, and many places in between.

A hard look at instream flow issues and a serious commitment to finding lasting water solutions are critical to the region's future. The political and legal approach to water is biased towards small steps and minor adjustments. Contemporary regional water decisions affect millions of people in the West. Much more is required than the status quo approach to these decisions, and new emphasis must be placed on problem solving and practical results. Slowly around the West, local communities, agencies, governments, and stakeholders are forging negotiated resolutions that apply water fairly to the greatest number of beneficial uses.

The purpose of this Chapter is to broadly summarize several instream flow issues on a state specific and regional scale, and make discrete recommendations for developing a new, more solutions-oriented approach to water problems in the West. This Chapter makes general comparisons between state issues and developments, but does not grade or rank individual states. These recommendations are solutions-based, and if acted on have the potential to produce lasting results to thorny water disputes whereby instream and fishery needs and consumptive water needs are met. The Chapter argues that the West's water future hangs in the balance, and concludes that rekindling a Western "pioneer-interest" in participatory democracy is a critical step towards ensuring a healthy water future.

The Water and Fish Baseline

Basic Water Overview

Water defines the West either by its abundance or its scarcity. California provides a useful illustration of this hydrological fact, where about seventy-five percent of rainfall occurs in the northern part of the state yet about seventy-five percent of the population lives in the southern part of the state. Water has always been one of law's most politically charged areas. Therefore, it is hardly surprising that "water litigation is a weed that flowers in the arid West."

Water is state property. Western states have detailed code and regulation governing water. While federal water law exists, and federal water projects move large quantities of water in the West, the administration of water use is largely left to the individual western states. State based water allocation decisions are crucial to instream flow. Each western state relies on the prior appropriation doctrine. A short set of uniform principles applies under that doctrine. These include: "first in time, first in right," beneficial use without waste, and "use it or lose it." The private property interest in water is a right to the advantage of the use of water. Most western states now recognise fish and wildlife as beneficial uses.

Basic Fish Overview

The presence of endemic fish species in the West is especially high. Sixty-nine percent of the native fish in the Colorado River Basin are found nowhere else in the world; in the intermountain Bonneville Basin that figure is forty-five percent, and in the Sacramento and San Joaquin basins in California it is thirty-three percent. The diversity that these species represent is irreplaceable.

They are not doing well. In the last century, at least twenty native fish in the West have become extinct. Salmon and steelhead are in equally dire straits. To take a state specific example, California faces a catastrophe in biodiversity and species loss. Among the fifty states, it ranks second in numbers of freshwater fish species that are declining. Add anadromous fish, and the state leads the nation in species loss and imperilment. At one time, California waters contained 116 native fish species. Today, sixty-two percent of those species are either extinct or will require serious actions to prevent extinction. A leading factor in aquatic and fish species decline is insufficient instream flow in the region's streams, creeks, and rivers.

The Intermountain West States of Colorado, Idaho, Utah, and Wyoming

This Chapter groups Colorado, Idaho, Utah and Wyoming together to analyse regionally similar developments and challenges.

1. Developments

   a. Grappling with Difficult Issues .

These intermountain west states apply traditional approaches to instream flow issues. However, in each state there is some reason for instream flow optimism. Colorado is the casebook example of a highly regimented approach to water, but it is now possible in that state to obtain "Recreational In-Channel Diversions." In Idaho, many stakeholders and officials are taking notice of watershed-specific partnerships like on the Lemhi River, which is home to a successful flow leasing program. In 2006, the Utah legislature passed a bill that reauthorised a legislative water issues task force, and directed that the force "shall review and may make recommendations on: (a) instream flow." In Wyoming, during the 2005 legislative session, the state senate overwhelmingly passed a municipal storage bill that would have allowed municipalities to use storage fights in existing reservoirs for nonconsumptive purposes such as fishery flows.

Water scholar Frank Trelease once Remarked

The law is a mechanism for getting things done, for accomplishing the purposes of society, for requiring some things and forbidding others. If the people of the United States or of a state desire to keep water in a stream or to put it back in a stream a law can be framed to do the job.

In the face of Colorado's worst drought in recorded history, state lawmakers put forward forty-three water bills, which experts observed was remarkable given the state's historical predilection towards turning a blind-eye to careful management of its water resources. The fact that these states are slowly but surely addressing difficult instream flow issues is a major reason for regional optimism. b. Recreation Is a New Driver

Agriculture has historically been the largest user of water in the West. In Idaho, for example, agriculture's presence is overwhelming and accounts for ninety-six percent of water withdrawals and ninety-nine percent of consumptive use. Despite this historical dominance, agricultural diversions peaked several decades ago. The "Lords of Yesterday" may rule the natural resources law of the land, but economies are changing around the region.

Recreation is a major new economic player. Healthy rivers and natural resources fuel recreation. In 2002, anglers spent approximately $423 million in Wyoming. Anglers spend approximately $46 million annually chasing wild trout on the Henry's Fork and South Fork of the Snake River in Idaho, and would pay $32 million more a year to catch more or bigger fish. The West is also experiencing rapid growth in urban whitewater recreational kayaking parks, and Colorado is at the forefront of this economic trend. Recreational instream flow is a wave of the future.

Challenges

Making the Most of Incremental Successes : Idaho's minimum instream flow laws protect less than one percent of the state's 93,000 stream miles, and are always junior rights, despite the state approving minimum instream flows in 1978. Even though Colorado adopted an instream flow program in the intermountain west, it was not until 2002 that the program was expanded to allow the Colorado Water Conservation Board to acquire existing rights to improve rather than just preserve the natural environment. In this region, only the states may hold instream rights and protect those rights.

Change takes time. For example, despite precedent-setting efforts to save Mono Lake in California, the Mono Lake Committee is still working to fully restore critical tributary streams almost thirty years after its landmark legal victories against the City of Los Angeles. Survival requires keeping your "eyes on the prize." It will remain a serious challenge in these particular intermountain states to find opportunities for instream flow benefit within such rigid systems. Those engaged in instream flow issues in these states should appreciate incremental success and certainly treat it as a victory, because in many regards a small instream flow improvement often means a large fishery or other natural resources benefit.

***Growth in the Face of Aridity* :** This challenge may be the largest gorilla in the room. As it was in the beginning, so shall it be in the end: the paramount question in the West is whether nature bounds growth. Western communities elected to follow at least the spirit, if not the literal meaning, of the maxim "rain follows the plow." Western history reflects blind devotion to the belief that technology always prevails over nature.

Cities in the region often defy common sense. Moreover, America's fastest growing states are located in the West. The facts speak for themselves. Consider Utah, which ranks among the driest states (second) and the fastest growing (seventh) in the nation. By 2030, it is estimated to add over another million new residents. Balanced against growth implications is the health of the state's 14,000 miles of perennial streams, and the habitat that those streams provide for Bonneville, Colorado River, Lahontan, and Yellowstone cutthroat trout and other native species.

California is an extreme example of growth and water use in the West. Californians use more water than any other state in the union. And, the state is expected to add approximately twelve million people almost fifty percent of its current population by 2030. This number is roughly equal to the current population of Pennsylvania.

The severity of drought conditions during the last several years in the intermountain west adds a complicating dynamic to this particular challenge. The 2002 drought in Colorado resulted in the lowest streamflows in the state in at least 100 years and probably in 500 years. Clearly, western communities can exist indeed thrive in western climatic conditions. However, growth management and therefore water management in the face of aridity and drought will remain a serious challenge for western communities. Fact must prevail over fiction.

Since 1969, it has been possible under Montana water law for water rights to be used to protect fish in the state's waters, making it in many ways a more progressive state than its intermountain neighbours discussed in this chapter.

Developments

***Water for Fish Becomes a Reality* :** Instream flow for fish is now an accepted part of life in Montana. Montana's approach to water leasing is the best example of this development, and one of the best success stories is the restoration work underway in the Blackfoot

River Basin. Partnerships between local stakeholders, irrigators, and three state and two federal agencies have resulted in a sustained and hugely successful water leasing program that has restored viable native fish populations and increased aquatic integrity. Innovative leasing solutions for instream flow can also assist in drought response.

Native bull trout and cutthroat trout make their home in the North Fork of the Blackfoot River in Montana. Drought and diversions can strand fish as they are migrating between the North Fork and the mainstem of the Blackfoot. The conservation organisation Trout Unlimited and local landowners negotiated a water leasing agreement that resulted in an additional 18.5 cubic feet per second (cfs) of water remaining instream in a typically dewatered section of the river in exchange for irrigation efficiency upgrades for the landowners including pumps, pipes, a centre-pivot irrigation system, and a solar-powered stock-watering well.

At the start of the twenty-first century, however, the future of Montana's water leasing program was uncertain. In 1989, the state legislature passed a leasing pilot project. Under the project, the state Department of Fish, Wildlife, and Parks (DFWP) could lease fights on four specific streams. Two legislative sessions later, in 1991, the legislature increased the number of eligible streams and lengthened the pilot to ten years. In 1995, the legislature passed an additional statute to allow temporary instream flow transactions between water users and private parties, and authorised the private leasing program until 2005, requiring additional legislative action to extend the program past that date. Then, in 1999, the legislature amended the private program to extend the lease period maximum up to thirty years for certain leases.

From 1989 to 1999, the DFWP completed ten leases. The political mood also changed dramatically over the 1990s. The debate started with fierce argument in the early 90s between instream and consumptive use interests, but began to diffuse in the mid 90s because of the success of locally based, collaboratively-endorsed pilot projects that were good for fish and good for farmers. By the end of the decade, the legislature had passed positive amendments to the leasing program. At the start of the 2005 legislative session, support from a broad range of interests rolled in for water leasing. In 2005, the Montana legislature passed, and Governor Schweitzer signed, House Bill 308, making the state's private water leasing program permanent.

Forward Looking Legal Results **:** In 2002, in the In re the Adjudication of the Existing Water Rights to the Use of All the Water, Both Surface and Underground, Within the Missouri River Drainage Area, Including All Tributaries of the Missouri River in Broadwater, Cascade, Jefferson, and Lewis & Clark Counties, Montana (Basin 411) (Bean Lake III) case, the Montana Supreme Court reaffirmed that fish, wildlife, and recreation are beneficial uses, and held that the prior appropriation doctrine is flexible enough to allow an instream right even without a physical diversion of water. According to the court, the "touchstone" in the analysis is the application of water to beneficial use. Thus, so long as water is put to beneficial use, diversion is unnecessary. In other states, overemphasis on diversion has limited instream legal advancements.

The Bean Lake III decision takes a common-sense approach to water. It also makes economic sense given that recreational angling in the state generates approximately $300 million annually. The vociferousness of opponents, however, indicates that water law evolution in Montana is risky business.

More recently, the Montana Supreme Court waded into groundwater issues in the state. In spring 2006, in Montana Trout Unlimited v. Montana Department of Natural Resources & Conservation, the Court held that the state Department of Natural Resources and Conservation's (DNRC) consideration of hydrologic connectivity between surface flow and groundwater use failed to "account for impacts to surface flow caused by the prestream capture of tributary groundwater." In the early 1970s, the Montana legislature revamped the state's administrative approach to its water rights system, including determining that certain basins were over appropriated. During that reorganisation, the state also enacted "a moratorium on new applications in the over-appropriated basins." The basin closures included watersheds in the upper Missouri Basin, and specifically the Smith River. The Smith River is home to a world class recreational trout fishery, and low instream flows have harmed that fishery.

Exceptions exist to the general closure law. The most notable exception is for new groundwater applications. To qualify for an exception, the new groundwater application cannot be "immediately or directly connected" to any surface water. Yet, despite the heavy lifting required of that phrase, the legislature failed to define "immediately or directly connected to surface water." As water users turned to groundwater to supplement limited surface water supply, the state agency's perspective on this phrase came under serious scrutiny from many corners.

In Montana Trout Unlimited, the court reviewed the DNRC's interpretation of this phrase. The court acknowledged the discretion vested in the agency to define and create methods to process applications, but ultimately ruled that the plain meaning of the basin closure law was to prohibit any processing or granting of groundwater applications that have "an immediate connection to surface flows or a direct connection to surface flows, or both." DNRC had identified two ways in which groundwater pumping affects streamflow in the Smith River Basin. DNRC identified the first way as prestream capture, which is when pumping intercepts groundwater that would otherwise become surface streamflow. DNRC called the second way induced infiltration, which is when pumping pulls surface water away from the stream towards the groundwater well. The agency's analysis showed that prestream capture is more harmful to surface flow than induced infiltration. DNRC, according to the court, had only considered one connection induced infiltration at the expense of another prestream capture, which the agency admitted caused direct impact.

The court noted that Montana water law was designed to protect both senior water rights holders and surface flows in the state's rivers. It is now clear that in the closed upper Missouri River Basin in Montana, state law prohibits granting or even processing groundwater applications where the end result of the groundwater use would be less surface flow. In Montana Trout Unlimited, the court correctly took an approach to the relationship between groundwater and surface water based on basic science and hydrologic fact.

Challenges: Building Alliances to Find and Implement Integrated Solutions

There may be no greater water issue in Montana than the need for uniform, programmatic treatment of groundwater use in "closed" basins in the state. Finding integrated solutions is a serious challenge. Alliances between unlikely partners can help find and implement solutions. Moreover, alliances are often needed between federal and state agencies who share authorities for the protection and restoration of aquatic species and water management.

Gallatin County, Montana, is an example of the pressing need to fend comprehensive and integrated solutions for groundwater management in the state. This county, located on the northern border of Yellowstone National Park, is Montana's fastest growing. Since 1986, the number of new groundwater appropriations has doubled. In December 2003, the Gallatin River had its lowest flow level in recorded history because of a prolonged drought and the increase in groundwater reliance.

Unique alliances have formed to file administrative protests to new groundwater appropriation applications. The subsequent administrative process has fostered citizen-based initiatives designed to address mitigation for groundwater impacts to surface water instream flow levels. These include the creation of a County Task Force and the Greater Gallatin Watershed Council. The risk and expense of contested administrative hearings have also spurred parties to reach innovative settlements.

The Big Hole River Basin in Montana is an example of proactive agency collaboration. The federal and state agency alliance in that basin should serve as a model for other parts of the state dealing with difficult instream flow issues. In the Big Hole, the Arctic grayling (Thymallus arcticus montanus) are struggling to survive. It is the last refuge for the grayling in the lower forty-eight states. Insufficient instream flow is a limiting factor. The United States Fish and Wildlife Service and Montana Fish, Wildlife, and Parks (FWP) are in the process of developing a multi agency, multi stakeholder, basin-wide fish restoration program that includes developing and implementing a strategy for improving instream flow for the grayling.

The rapid growth in the greater Yellowstone area underscores the necessity for integrated approaches to groundwater use. The Big Hole example concerning the future of the Artic grayling highlights the need for agencies to be willing and able to work across agency boundaries. Sound hydrologic understanding should form the basis for those solutions. Strong alliances between instream flow advocates and agricultural communities should be formed to develop solutions.

Water management, water law, and water policy in California are more complex than in any other place in the West. California, however, may be the envy of instream flow advocates in other states.

The Public Trust Doctrine : In 1983, the California Supreme Court determined in National Audubon Society v. Superior Court (National Audubon) that the state as a sovereign entity has the authority and the duty "to protect the people's common heritage of streams, lakes, marshlands and tidelands." This power and responsibility exist sui generis. Los Angeles Department of Water and Power (DWP) had been diverting almost all surface flow from four out of the five tributary streams of Mono Lake to supply the city with water. This diversion was occurring pursuant to rights granted in the 1940s, and had significantly impacted the lake. In National Audubon, the Court held that the state water agency (State Water Resources Control Board) should reconsider DWP's water rights based on public trust concerns. While relatively few judicial decisions have followed National Audubon, the State Water Resources Control Board (SWRCB) has further developed and defined a powerful public trust role in water allocation in California.

Charles Wilkinson wrote that "[t]he recognition of the public trust doctrine in water law is the single strongest statement that historic uses must accommodate modern needs." This doctrine expressly requires consideration of public trust resources and instream flow needs in water planning and allocation based on twenty-first century knowledge. California is the birthplace of the public trust doctrine in water law. No other case in water or environmental law has generated as much buzz or sustained as long a movement for the restoration and protection of a single place Mono Lake than this one.

State Water Resources Control Board : "The history of California is written on its waters from the Eel River, to the Salton Sea, to the Colorado River, to Lake Tahoe," and in a perfectly Californian way this history has even made it to the bright lights of Hollywood. Today, the chief architect of California's water future concerning instream flows is SWRCB. SWRCB is a state agency with hundreds of employees, an operating annual budget of hundreds of millions of dollars, and a five member board. Consideration of fish and wildlife preservation and enhancement is an overarching statutory requirement for SWRCB. Indeed, SWRCB instream flow determinations are a condition precedent to issuance of new water rights.

In 1983, the California Supreme Court concluded in National Audubon that: "[T]he function of the Water Board has steadily evolved from the narrow role of deciding priorities between competing appropriators to the charge of comprehensive planning and allocation of waters." Great opportunity comes with this mandate. SWRCB can undertake regional, integrated water planning efforts to meet this responsibility regarding instream flow. For example, in the 2004 legislative session, Governor Schwarzenegger signed Assembly Bill 2121, requiring that SWRCB develop instream flow guidelines from San Francisco Bay north to the Mattole River, Mendocino County, which is an area that covers Marin, Napa, Sonoma, Mendocino, and southern Humboldt counties. The geographic scope of this regional instream flow effort is larger than many northeastern states.

SWRCB has not ignored opportunities to expand its authority to protect and enhance instream flow. Very recently, the California courts provided additional impetus for SWRCB to act. Countless scholars have analysed SWRCB's ongoing efforts to balance its adjudicatory role regarding water quality objectives and water rights in the San Francisco Bay-Sacramento-San Joaquin Delta. In February 2006, the Third Appellate District for California determined that SWRCB's "power to subject appropriative water rights to terms and conditions necessary to carry out water quality control plans" is more mandatory than permissive.

Saltwater intrusion is the major water quality limitation in the Bay-Delta. Freshwater outflow helps address this limitation. SWRCB Decision 1641, in part, sought to assign water quality objectives to water rights holders, including to the federal Central Valley Project and the State Water Project. Multiple parties objected to the allocation of responsibility and appealed.

In State Water Resources Control Board Cases (Robie), the court determined that:

> *when a water quality control plan calls for a particular flow objective to be achieved by allocating responsibility to meet that objective in a water fights proceeding, and the plan does not provide for any alternate, experimental flow objective to be met on an interim basis, the decision in that water rights proceeding must fully implement the flow objective provided for in the plan.*

Because SWRCB had clearly identified in its 1995 water quality plan that the subsequent water rights proceeding would serve as the enforcement mechanism to ensure attainment of water quality objectives, it was required to exercise this power "for the public at large" and apply terms and conditions to the subject water rights in order to comply with its own water quality plan. The result is the merging of quality and quantity before SWRCB in water rights proceedings. More legal analysis of this decision is surely on the way.

However, this much is clear: The Robie decision will greatly influence the next generation of water rights disputes in California and the activities of SWRCB on the instream flow front. c. Water Transfers and Markets

California's water code is one of the most advanced for instream flow transfers and markets. California is unlike Montana, which only has a leasing program.

It is also unlike other intermountain west states, which allow only the state to hold such rights. In California, any person or entity may own or dedicate an existing water right to instream beneficial use for fish and wildlife and recreation benefit in perpetuity. The one restriction is a prohibition on dedications of new rights. The transfer and dedication of existing rights to instream use effectively creates a legal work-around that avoids the precondition of physical possession of water for obtaining a new right.

Challenges

Groundwater : The existing regulatory approach to groundwater in California overlooks basic science. SWRCB believes that it may only take permitting authority over groundwater use after a series of complicated physical criteria tests are met. The practical effect is that SWRCB takes jurisdiction less often, and groundwater use exists relatively unregulated compared to surface water use. The most troubling aspects of this particular challenge are the lack of political will and the complete absence of a forward-thinking consensus for revisiting the issue. These failures persist despite one of the state's leading water experts providing a solutions-roadmap at SWRCB's request.

In 1977, Governor Brown of California convened what became known as the Governor's Commission to Review California Water Rights Law. The commission's final report underscored that "groundwater is subject to what has been called the 'tragedy of the commons.'" Almost thirty years have passed since this landmark report, and little has changed.

California should develop a comprehensive approach to groundwater that effectively and fairly blends local needs with a comprehensive state perspective. California has taken an atypical "cannot do" approach to this important issue, which is all the more surprising because the experts are in consensus regarding the need to fix this regulatory problem.

Where Is the Market? : A "quiet revolution" for instream flow took place in California in 1992 when the state legislature amended the water code with section 1707. This section is one of the West's most advanced water code provisions for instream transactions. Unfortunately, it is very likely that the number of section 1707 transactions in California can be counted on two hands. Instream flow advocates in California are correct to ask: Where is this market?

Administrative processing hurdles and cost are two limitations to an active section 1707 market. Because section 1707 is styled as an instream transfer, rigorous transfer standards apply, including the "no-injury" rule. Moreover, because so few section 1707 transfers have occurred, inter and intra agency procedures are Convoluted and not easily understood, which of course increases willing transferors' and transferees' transaction costs.

A bigger obstacle is California's fascination with all things large. The state's water politics have always been preoccupied with the movement of water from north to south. Indeed, the federal Central Valley Project and the State Water Project are designed to facilitate transfers of huge quantities of water. Institutional structures like the Environmental Water Account also emphasize trading large quantities of water in the Delta and Central Valley. However, a drastically different scale exists. On that scale, a transfer of ten cfs or less can make the crucial difference in instream flow for fishery or other natural resource needs. This scale is unfortunately too often overlooked to the detriment of instream flow opportunities and the natural resources reliant on such flow. Section 1707 is a perfectly placed legal tool to correct this oversight.

Living up to Potential: Permitting Systems that Work for All Stakeholders : California's regional communities and water stakeholders are faced with the challenge of fixing the state's water rights permitting system. This challenge is playing out north of San Francisco where there are approximately six hundred or more pending applications for new appropriative water rights and surface diversions, mostly in the Russian River basin, in Sonoma and Mendocino counties. Many of these applications have been pending ten or more years. In some watersheds where applications are pending, which are also watersheds for endangered or threatened anadromous fish, the level of unauthorised diversion is as high as seventy-seven percent. In some cases, the unauthorised diversion has been ongoing for decades.

The California Legislative Analyst's Office recently concluded that:

> *existing funding levels allow the [water board] to process around 150 applications annually. However, the [water board] currently has a backlog of over 680 pending applications. Even with no new applications for permits, it would take over four years to process all of the backlogged applications at the current rate. Currently over 1,000 permittees are waiting to be inspected and licensed. In addition, staff inspect about 120 water fights annually at current funding levels. This reflects annual monitoring of less than 1 percent of the water rights under the [water board's] enforcement jurisdiction.*

There is no dispute that the "orderly and efficient administration of the water resources of the state" is in the best interest of all stakeholders. This challenge creates a clear choice for Californians. In this region, stakeholders can either look the other way while a wild, wild west approach to water continues, or they can take action and develop consensus solutions for an effective and timely permitting system that works for all beneficial uses, including monitoring, compliance, and enforcement. The problems will only worsen in the absence of common-sense, good government reforms to the water permitting system in northern California.

A New Water Future

Relationships Matter When it Comes to Water

Aldo Leopold wrote that "[t]here are two things that interest me, the relationship of people to their landscape and of people to each other." In the West, it may be more important to scrutinize the relationship of people to water and of people to each other about water. At the end of the day, law, policy, and science matter far less than relationships when water is at stake.

The Klamath River, for example, was once the third largest salmon producing river in the West, behind the Sacramento and Columbia River systems, and it is the West's third largest river. The ecosystem diversity of the Klamath Basin is extraordinary. The cultural and social diversity is remarkable. Native American peoples have lived in the area for more than ten thousand years. The farming community is a strong presence in the basin, and proudly identifies with the generational and lifestyle choice of farming the land. The commercial fishing industry in northern California and southern Oregon plays an important role in Klamath River issues. And, the conservation and environmental organisation community is deeply involved in the basin.

Conflict has been the norm in this basin. These multiple interests have continually clashed. The result is a cycle of conflict similar to the ecological principle of competitive exclusion. This cycle exists because stakeholders (knowingly and unknowingly) have fostered a controversy whose dominate trait is a battle for exclusive right over a scarce resource. Relationships are the key to ending this water war. Only when these parties start listening to each other and working with each other for a common good peace on the fiver will the necessary space emerge for them to problem solve these tricky water issues in a way that produces viable solutions. No other approach holds any promise for lasting outcomes in the Klamath.

It is difficult to overestimate the importance of building bridges and partnerships. Yet, it is very easy to put off undertaking the time-intensive due diligence to build authentic relationships. It takes time. It is difficult. Healthy and successful relationships require stakeholders to engage in give-and-take. The most important prerequisites, however, are the ability and willingness to listen and learn from those who hold different views and opinions.

The rewards from these non-traditional relationships and partnerships for instream benefits are potentially huge. In 2005, Utah ranchers visited successful water leasing projects in Montana and had the opportunity to listen to their Montana counterparts' experiences with leasing. The exchange visit generated favourable press coverage in Utah. The Provo Daily Herald called the Montana effort a "groundbreaking program that has allowed farmers and ranchers to earn money for unused water while benefiting wildlife," and quoted one participant in the exchange, who offered his opinion that "where it is a willing-seller, willing-buyer situation, those things should be allowed." The first step is a united dialogue. The parallel step is to insist on civility.

Finding solutions is tough, but in this regard our mothers and fathers and elders were absolutely right treat people with respect and just as you would want them to act in return. There is simply no justifiable reason for the scale and intensity of hatred in water issues around the West. All of that misguided energy spent on hostility drives stakeholders farther apart, and would be much better directed towards problem solving. We are all in this together like it or not. Let us collectively check the vitriolic gamesmanship that takes place in the public marketplace at the door. Water stakeholders must self-initiate a united, relationship-based dialogue on a regional scale and in individual cases. They must ensure the spirit and tone of this dialogue does not make us feel embarrassed because of the vulgarity and mean-temperedness, but rather makes us feel proud to know, live, and work beside each other in our regional communities.

Instream Flow and Regulatory Certainty Can Go Hand in Hand

In Montana, one of the biggest challenges facing water stakeholders is increased pressure to meet surface water limitations with groundwater use in closed basins. This challenge puts fishery and aquatic resources at risk and threatens senior water rights holders. In California, regulatory uncertainty associated with the permitting of surface water rights applications north of San Francisco produces extreme business risk for the regulated community. Indeed, not knowing whether, when, and how SWRCB will process an application for appropriative right greatly undermines business planning. Moreover, in some instances, existing senior water rights holders suffer the consequences of competitors gaining an unfair business advantage through unauthorised diversion. This situation also directly impacts the recovery of coho salmon and steelhead in coastal California streams.

Instream flow protection and regulatory certainty can, and should, occur simultaneously and in conjunction. For example, Gallatin County is now home to many emerging reforms that require mitigation for groundwater impacts to surface water instream flow levels. In that basin, unlikely alliances between fishery advocates and senior water rights holders helped develop these ideas. Conservation organisations, state resources agencies, private owners of hydroelectric dams, and surface water irrigator associations joined forces in protest of new pumping permits. Together they worked to address groundwater-surface water connections and impacts so that instream needs and senior water rights holders were adequately protected. In the California example, a broad and diverse working group including Trout Unlimited has emerged for the purpose of working together to create a more timely and effective permitting system that provides regulatory certainty for applicants and protects fishery needs. A potential consequence of such a collaboratively-created system could be the diminished need for parties to protest future applications.

Instream flow advocates and water users have let the gulf between them widen too far. Better relationships between instream flow advocates and water users should daylight creative options for solving problems. Finding solutions is tough, but satisfaction of instream flow needs and water rights certainty can fit together just as nicely as apple pie and baseball. Do not let anyone convince you otherwise. Water stakeholders must understand and embrace this positive dynamic, and then build solutions that sustain and enhance the compatibility between these seemingly disparate interests.

We Need More Physical Solutions

For far too long, water stakeholders in the West have argued over whose water use is more important. Arguments framed that way guarantee losers. Conflict will rule the West's water future as long as stakeholders continue to mistakenly believe that they are required to choose between mutually exclusive outcomes. It is imprudent to allow rigidity to control solutions. The discussion does not have to always be about fish or farms, people or the environment.

Stakeholders need to reframe the questions facing them. A world of opportunity awaits those stakeholders who ask and answer how to best manage and optimise the scare resource of water to maximise multiple interests and beneficial uses. It can be about fish and farms, people and the environment.

The physical solution doctrine is the key to this refraining. Although the doctrine is based on principles of equity, scarcity of water is the contemporary motivator. The scarcity of water and its preciousness require critical and collaborative thinking to put it to maximum use for the most purposes.

Investing in a physical solution approach means stakeholders will undertake critical analysis to locate and exploit opportunities for mutual gain. As Gregory Thomas puts it, "[p]hysical solutions are the devices crafted by the courts in California to reduce conflicts among competing water users." Mechanisms to achieve such reconciliation include conservation and efficiency strategies, reservoir re-operation, delivery adjustments, groundwater banking and conjunctive use programs, and transfers. A physical solution is to a water allocation challenge like high-speed internet is to information exchange. Both rely on expanding options and capacity to manage a resource (water or data) across the broadest range possible.

The Yuba Accord in California is a good example. The Yuba River originates on the Sierra Nevada Range's west slope and drains approximately 1,339 square miles of Sierra, Placer, Yuba, and Nevada counties. The Yuba flows into the Feather River below Marysville, California, and the Feather in turn is a tributary to the Sacramento River. The lower Yuba River wild chinook population is one of the most significant remaining populations in the entire Central Valley and is also home to one of the last naturally serf-sustaining steelhead populations.

The Yuba River has been at the centre of one of California's longest-running water disputes. In short, the California SWRCB exercised its continuing authority over Yuba County Water Agency's (YCWA) water rights in 2001 and revised those rights to require more water instream for fishery resources in the lower Yuba River in SWRCB Decision 1644. That decision represented the culmination of approximately fifteen years of controversy. During subsequent state court litigation challenging Decision 1644, parties negotiated a complex proposed settlement agreement called: The Proposed Lower Yuba River Accord.

The hallmark of the proposed settlement is the physical solution doctrine whereby a fisheries agreement would establish higher instream flows during periods of the year; a water purchase agreement would create long-term purchase agreements between YCWA and the United States Bureau of Reclamation and California Department of Water Resources; and, a conjunctive use agreement would establish a comprehensive program between YCWA and its customers to integrate surface water and groundwater resources. Seventeen parties are involved in the integrated agreements, and the accord would optimise operations and water use in the Yuba River system to significantly increase instream flow, produce the first major long-term water acquisition for the state's Environmental Water Account, and generate revenue for YCWA to implement flood control improvements. The physical solution approach helped parties broker the proposed end to this long-running water dispute.

Finding solutions is tough, but allowing the frame of the allocation discussion to remain only about competition rather than cooperation will slow instream flow efforts going forward. It does not have to always be about winning or losing. The most short-sighted definition of success is beating the other side. Water stakeholders must unpack and then repack the community dialogue so that the operative spirit is collective problem-solving. A physical solution perspective can help with this necessary shift.

Doing Nothing is Unacceptable

Water stakeholders must be proactive. Doing nothing about water issues will not cut it anymore. Failure to find lasting, community-supported solutions to the West's water challenges will substantially affect the economy and environment of the entire region. The challenges will only worsen in scale and complexity if not addressed now.

One precise example previously discussed in this chapter proves that inaction is simply untenable. In northern California in 1990, the number of pending applications for new appropriative water fights and surface water diversion was relatively small. In 1997, at the time California's State Water Board's Staff Report on these issues was published, the number was approximately eighty-one for the Russian River watershed. In 2004 when Trout Unlimited and the Peregrine Chapter of the Audubon Society filed an administrative petition with SWRCB requesting solutions for these problems, there were over 276 pending applications in just five counties.

It would have been far cheaper and easier to address the problems associated with eighty-one pending applications than with 276. This, of course, just definitively proves Marc Reisner's astute point that "in the West, of course, where water is concerned, logic and reason have never figured prominently in the scheme of things." Today, the number of pending applications is approximately 600.

So much more is needed than business as usual for the regional and state water issues and challenges discussed in this chapter. The status quo cannot carry the day on water issues in the West. Finding solutions is tough, but if you listen carefully to the success stories in this Chapter and to other similar stories, a movement is on the horizon and an opportunity is coming around the corner that may only come once. Do not let it pass.

When called to serve their country, Americans stand up. When challenged to go to the moon, Americans responded. Westerners must stand up and respond now because our collective water future is on the line. Stakeholders have to take the scary step together and rush in where others have feared to tread in the past, if for no other than reason that the alternatives are far more grim. Water stakeholders must find the willingness to grapple with these unquestionably thorny issues concerning water management, water law, and water policy in the West. Step up to the plate and take action. Time is wasting. The challenges are not going away.

Winston Churchill is widely considered to have remarked that "Americans always can be counted on to do the right thing, after they have tried everything else." Either stakeholders across the West come together to determine a shared water future or they do not. We have already tried everything else. The lives of millions of people and the region's remarkable natural resources hang on this pivot point.

The West's water future lays exposed in the balance. The fact of the matter is like a sharp blow to the gut: none of us can live without water. Rekindling a Western, pioneer interest in participatory democracy is a critical step towards ensuring a healthy water future. The scholar Donald Snow notes that observers have consistently applauded Americans' ability to work together and "individual willingness to engage in local problem-solving and political action." Powell's watershed commonwealths conceptualised a "participatory democracy" for natural resources decisions. The West's legacy is one of people who faced adversity both pragmatically and head on. Democracy requires involvement not passivism. It is a direct contact sport. We need to rekindle a pioneer-like spirit, which I define as active community involvement to overcome adversity together. We need to then use this spirit to chart a new and responsible water future for the West. A review of instream flow issues and water management, water law, and water policy around the West uncovers many challenges. Some of the challenges, like drought, have been around since the beginning. Others, like advancing water markets, are relatively new on the scene. It comes down to all of us right here and right now. One by one, individual by individual, community by community, a reapplied belief in the Western ability to solve tough problems can translate to renewed commitment. We must rescue ourselves from ourselves.

Coming together to discuss water issues is true to Powell's vision of watershed commonwealths. This concept implies joint ownership and joint effort derived from common hydrological circumstances. It suggests a commitment to a common allegiance of living and working together because of those shared circumstances. Through a civil discourse and sitting down together we may discover how to conduct ourselves and in the process find a shared "lived ethics." This "ethic" is a "kind of community instinct in-the-malting." Water is an appropriate resource around which to build such a community ethic.

The West is a mythical place. It oozes confidence and optimism. There is much to be excited about from an instream flow perspective. Stakeholders can come together and forge resolutions that apply water to the greatest number of beneficial uses in a fair way. Physical solutions prove that point. So too do the many success stories described in this chapter. A new water futurc can be charted, but it is entirely up to us. Only if we take collective action now to foster public dialogue and decision making on the water issues in the West will we find lasting solutions that manage water for the greatest good. Finding those solutions also holds the remarkable promise of finding peace with each other over water at the same time. Those solutions are our best chance to ensure healthy natural systems and communities, and finally produce a society that does justice to the region's natural beauty and people.

Water Rights Law in Theory and Practice

This Comment explores streamflow protection issues as they intersect and conflict with existing water regulation practice in the state of Oregon. Numerous factors make the protection of streamflows for ecological benefit difficult. These factors include: the practical limitations of managing a complex and unpredictable natural system; the administrative reluctance stemming from local opposition to non-consumptive uses and years of enforcing the status quo; and the formal, legal opposition on the part of those who fear that protecting streamflows will interfere with their own rights. The difficulties represent not only conscious choices but inhere in the antiquated doctrines of Western water law. This Comment explores particular impediments to streamflow protection raised by the prohibition against legal injury to others' water rights. By illuminating the practical, regulatory, and legal parameters of injury analysis and superimposing this framework on the realities of Oregon's administrative and legal regulation of streamflow, this Comment seeks to assist parties in navigating the complexities of instream transfers. Outside of Oregon's state lines, the hope is that this analysis will aid parties in other western states and the states themselves, as they develop streamflow protection laws and policies.

The Comment begins with an overview of Oregon's streamflow protection laws and the administrative and regulatory processes designed to apply the law on the ground. Next, the prohibition against injury to existing water rights is examined in detail. With this background in mind, the practical, administrative, and legal challenges to implementing streamflow protection are discussed. Central to this Comment is the analysis of these challenges and the questions they raise are they based in sound law and policy? Do they treat streamflow protection on an equal legal footing with other types of water use? What can be learned from a close examination of these hurdles? Finally, the Comment concludes with a discussion of successful models of streamflow protection and a summary of recommendations for the future.

Oregon Water Law

Because water in the arid West is a limited resource, a system for distributing it during times of shortage has become an entrenched part of Western law. This law is called the "prior appropriation doctrine." The most basic tenet of the doctrine is that the first person to appropriate water from a source has the most senior right and will therefore be the last appropriator cut off during times of shortage. Conversely, later appropriators, called juniors, will be required either to shut off or diminish the amount of their diversions dung shortages to leave water in the river for the seniors.

Instream Water Rights

Oregon was one of the first states in the West to protect water instream when it removed streams feeding waterfalls in the Columbia River Gorge from appropriation to protect the streams' "scenic beauty." This step was followed by the establishment of minimum perennial stream flows in 1955. Minimum flow protection had limited effect, however, because the flows took precedence only over appropriations later than 1955. In 1987, the Oregon legislature took the final step toward meaningful instream protection by creating a legal right to maintain streamflows. The Instream Water Rights Act (the Act) begins by declaring that instream uses are beneficial uses and have an equal legal footing with "any other water right for which a certificate has been issued."

Instream rights are created in several ways. First, the Act converted the minimum perennial stream flows set in 1955 to instream rights with a priority date according to the date the minimum flow was established. Because of the late priority dates of these rights, however, their importance is minimal. Instream rights can also be created by state agencies.

The Oregon departments of Fish and Wildlife, Environmental Quality, and Parks and Recreation are empowered to request instream rights for public use. Again, these rights post-date the 1987 Act and are therefore at the back of the prior appropriation line. The only method of creating senior instream rights, and therefore the most important method, is the purchase, lease, or gift of a water right by a private party for conversion to instream use. Existing water rights of any seniority may be leased for temporary conversion to an instream use or may be sold for permanent conversion. Short-term leases are by far the most popular of these options and are therefore the primary vehicle in Oregon for creating instream rights.

Short-term lease agreements are not to exceed five years. These agreements allow landowners, who are unwilling to transfer their rights instream permanently, the freedom to do so for a trial period. In addition, landowners who are not using water, and are therefore at risk of forfeiting their right, can lease the right instream, stopping the clock on forfeiture by putting their water to "use." Instream lease agreements also enjoy the freedom to split the season between instream and out-of-stream uses.

On the Ground Regulation of the Right to Use Water in Oregon

In practical terms, the "first in time, first in right" priority system is difficult to administer. Two primary impediments to meaningful regulation of water rights are a lack of accurate measurement of diversions and a lack of resources to direct towards enforcement efforts.

State employees called watermasters, whom the director of the Oregon Water Resources Department (OWRD) appoints, carry out on-the-ground regulation of water rights in Oregon. Oregon is divided into twenty-one districts roughly following watershed boundaries. Watermasters and their staff are charged with the general duty of distributing water among users according to their rights. As part of their duty to oversee the distribution of water, watermasters are also charged with preventing wasteful practices. However, actions by the state of Oregon against wasteful water users are rare and only occur in extreme instances of waste.

The Mechanics of Leaving Water Instream

Maintaining flows instream requires a watermaster to regulate i.e. reduce or shut off completely the point of diversion (POD) of out-of-stream appropriators junior to the instream right, thereby allowing the proper amount of water to flow in the river. Though simple in theory, achieving instream flow requires careful planning and intimate knowledge of the river system. Instream rights are defined and protected as either points or reaches. When an instream flow is protected at a point, the watermaster is charged with ensuring that at least as much water as is called for by the instream right passes that point. The preference, however, is that instream rights be defined as "fiver reaches." In general, a senior instream right that has been acquired from an existing user and that is protected as a reach is maintained from the old POD where the landowner once diverted water out of the river to the mouth of the river. Instream rights can also be separated into several reaches, each with a different instream rate, to mitigate against injury to other users or to account for natural conveyance losses. If mitigation for loss of return flow is necessary, the instream right will be reduced by the amount of return flow at the point where return flow previously entered the stream. If this point cannot be identified, the right will be reduced at the old POD. Alternatively, if a river passes through a reach where natural flow losses occur by seepage from the river bottom, the instream right will be reduced to reflect this natural decrease.

The preceding discussion highlights the importance state regulators place on protecting existing rights from interference by instream rights. Before an instream right is created, either by request from a state agency or by conversion from an existing right, OWRD analyses the proposal for potential injury to other users. The doctrine controlling this review is called the "no injury" rule. It has deep roots in Western water law and is the forum for a majority of the challenges faced by applicants for streamflow protection. This comment focuses on application of the doctrine in the context of an existing user who wishes to transfer their right instream.

The "No Injury Rule"

Under the no injury rule, junior appropriators are protected by maintenance of the stream conditions that existed at the time of their appropriation. These conditions are not limited to natural conditions but include anthropogenic stream influences such as water returning to the river from an irrigated field. Understanding the rule requires a step-by-step explanation beginning with a description of the nature of the property right in water, a discussion of the context in which the rule is applied, and, finally, an in-depth discussion of the legal and practical contours of the rule itself.

The Property Right in Water

Once a water right has been issued by OWRD, the right to use the water becomes a fully vested property interest. Like an owner of real property, appropriators are entitled to enjoy their right free from injury by other property owners. However, the comparison to real property ownership only goes so far because a water right, while a vested property right, is limited to a right of use. "Ownership" of Oregon waters lies with the people of Oregon. The no injury rule therefore preserves juniors' use of water rather than their ownership.

Context for Application of the Rule in Oregon: Transfer Proceedings

Application of the no injury rule is most common in proceedings called "transfers" where decisions can be made to change from one water use to another or from a specific place of use to a new location such as in-stream.

The key question in transfers is how much of the water right may be transferred. Oregon is called a "paper right" state because transfers begin with the assumption that the entire certificated amount is transferable. Other western states define water subject to transfer based on historical patterns of use. In these states, a user historically used less than their allotted right, that pattern of use defines the permissible future use of the right.

Despite the "paper" status of Oregon water rights, appropriators in Oregon can nonetheless lose some or all of their water right through historical non-use. Non-use of water for a period of five or more years creates a rebuttable presumption of forfeiture. However, Oregon law exempts from forfeiture any water right under the following circumstance:

> *if the owner of a water right uses less water to accomplish the beneficial use allowed by the right, the right is not subject to forfeiture so long as (a) [t]he user has a facility capable of handling the entire rate and duty authorised under the right; and (b) [t]he user is otherwise ready, willing and able to make hill use of the right.*

In other words, no forfeiture results from using less than the "paper right" if the user is ready, willing, and able to use the full amount. The critical impact of this law is that downstream juniors, who historically depended on the unused portion of an upstream senior's right, are not legally entitled to that water if the senior meets the "ready, willing, and able" requirement. Under this doctrine, the face of the water right certificate is the first place to look at the beginning of a transfer proceeding. The parties then must work from this amount to determine how much water can be transferred without injury.

Defining Legal Injury

OWRD bars transfers when there is a determination that the change will injure other water users. In Oregon, injury occurs when an "existing water right [is] not receiving previously available water to which it is legally entitled." Another species of legal injury, enlargement, occurs when a water right is expanded beyond what is set out in the certificate. The crux of the injury analysis is determining what water was "previously available" and, thereby, what water juniors are legally entitled to. The question of what water was "previously available" is decided using the doctrine of "return flow." The concept of return flow is often confused with similar terms and physical phenomenon. Understanding the doctrine therefore requires familiarity with both legal and scientific concepts of return flow.

Contours of the Rule Part I: Legal Definitions of Return Flow

Oregon courts have adopted the definition of return flow as water that returns to the natural course of the stream from which it was taken after being applied by an appropriator.

"Possession" of water lasts until it "wastes back or percolates from lands or leaves the control of the owner of such lands." Stated differently, discharging excess irrigation water back into its source after use, absent the intention of recapture or reuse, effects an abandonment. Downstream users are then free to appropriate the returned water.

Two similar situations confuse the doctrine. First, ff the appropriator recaptures the water on her land before it rejoins its source, the water does not become "return flow." The appropriator has not lost control nor effectively abandoned the water. Second, ff the water finds its way onto adjacent lands or into a ditch or stream other than its original source after application, the "return flow" label does not attach. Appropriators who rely on this water, called "seepage" or "waste water," cannot compel continuation of the irrigation practice that results in the seepage. The key difference in this scenario is that the water has yet to commingle with its natural source and is therefore not legally appropriable.

Once defined as return flow, however, water is appropriable by downstream water users and "can not afterward be taken out by the original appropriator to the injury of other rights which have attached and vested to it." In other words, an upstream senior may not deprive downstream juniors of return flow by changing the use of water on her land or by changing the point of return of used water to the river below the downstream junior's point of diversion. As is often the case, using broad legal definitions risks oversimplifying reality. In the field, return flows are difficult to identify and differentiate from "seepage" or water destined for recapture and even more difficult to quantify.

Hydrologic Definitions of Return Flow

For the purpose of understanding return flow in the context of injury analysis, the most important hydrologic processes are those that remove water from the system "consuming" water or that result in water returning to its source and becoming return flow. The movement of water through soil after being applied for irrigation is controlled by two processes known as infiltration and redistribution. Infiltration is the process whereby water percolates into the soil. Redistribution refers to the movement of water once it has entered the soil. The rate at which water infiltrates from the surface into the soil depends on many factors. These factors include the rate of application of irrigation water, the hydraulic conductivity of the soil, the degree of saturation of the soil, the inclination and roughness of the soil surface, the chemical characteristics of the soil surface, and the physical and chemical properties of water. Once water infiltrates the soil, redistribution is controlled by gravity, pressure, and evapotranspiration from vegetation. On an irrigated field, evapotranspiration is the most significant redistribution force and is particularly important in the return flow analysis because it represents the amount of water that does not become return flow.

Evapotranspiration describes a number of processes by which liquid or solid water becomes water vapour. Among these processes, transpiration is particularly important. Transpiration is the process by which water is taken from the soil by plants, moved through the vascular system of the plant and evaporated into the atmosphere. Predictions of the rate at which transpiration occurs in crops are an important tool irrigators use to determine how much water to apply to their field.

Water transpired by an irrigated field represents water that will not become return flow water that is "consumed," in other words. Watermasters often rely on this figure alone to calculate return flow. They subtract "consumptive use" from the total amount of water the irrigator is diverting and label the difference return flow. The assumption is that all water not consumed returns to the river and is used by downstream water rights. This oversimplifies the analysis however, because redistribution processes other than transpiration also result in consumption of water. For instance, water flowing through an unlined, dirt irrigation ditch can escape the system either through surface evaporation or through infiltration and redistribution into the groundwater system.

Some water applied to the field and not consumed by plants or surface evaporation percolates into the ground and mingles with groundwater. Most groundwater is connected to a surface water source such as a river or lake. This hydraulic connectivity between ground and surface water is the basis of return flow. However, many variables control the amount and timing of groundwater movement into surface bodies. In discussing return flow and injury, therefore, timing of return flow is a critical factor.

Timing varies depending on the hydraulic conductivity, hydraulic gradient, and flow direction of the soil and groundwater at issue. Hydraulic conductivity is the rate at which water moves through the soil and varies depending on the soil-grain size. The hydraulic gradient describes the topographical inclination of the soil either toward or away from the source stream. Groundwater flow direction depends on the contours of the water table surface and affects where return flow may occur at a given time. In concrete terms, timing of return flow to a stream from an irrigated field is a function of the field's soil type and its topography with relation to the stream.

From the perspective of hydrology, the presence of return flow depends on a host of factors including consumptive use (transpiration) by crops and other mechanisms that transport water out of the system or back to the source. Imposed onto these factors is the additional variable of timing. The injury analysis should therefore take into account all of these factors to ensure that an accurate picture of return flow is drawn.

Obstacles to Instream Flow Protection

Obstacles to instream protection can be separated into three categories. First, practical limitations stemming from a lack of resources and local opposition to instream protection lead to deficiencies in the physical ability to manage the system. Second, "creative" administration of the water laws by OWRD often imposes an unfair double standard on instream transfers. Finally, the formal transfer process provides a forum for official challenges to instream transfers.

Practical Limitations

The first category of limitations on instream transfers are those imposed by a lack of resources directed to state administration of water and those imposed by political opposition to the idea of instream water rights. These are inextricably related. Opposition to instream protection comes from many who believe water should be put to "productive" uses such as farming and ranching. Often, those charged with administering instream rights share some of this opposition. Watermasters live and work in the agricultural communities they regulate, making impartial treatment of instream rights difficult at best and a conflict of interests at worst. In addition, regulation of water rights is complaint driven. As holders of instream rights, the state does not "complain" on behalf of instream rights with the same vigour a private right holder would. This problem is compounded by a lack of accurate knowledge about how much water is being diverted from a given river at any given time and by the sheer volume of administration required to oversee the system.

Individual watermasters oversee large land areas with hundreds, possibly thousands, of individual water rights. The John Day Basin for example, is roughly the size of Massachusetts and has only one watermaster. Monitoring each individual water right throughout the state is therefore impossible, and the system relies to a large degree on self-policing and complaints from right holders when their rights are infringed. Compounding this difficulty, a tiny fraction of diversions among those 70,000 rights are metred. In other words, measurement of water use is not mandatory. At the discretion of OWRD, water users may be required to install measuring devices on their diversions. As of 2003, however, only 8% of combined surface and ground water rights are required to do so. Governmental entities, including state and federal agencies, local governments, irrigation districts and water conservation districts which appropriate water are required to submit an annual water use report to the OWRD. Compliance with this rule is not 100%. Installation of a more comprehensive, statewise measurement program is an expensive proposition and one for which no budget currently exists. The importance of this lack of measurement must not be overlooked. A system designed to regulate water use and continually balance use among competing interests that has no means of measuring and monitoring such use is flawed. Without measurement, water resource management is, at best, a guessing game.

The practical problem of doing a big job with insufficient people and information is compounded by the complaint-driven nature of Oregon water law. Water users commonly divert more than their certificated right, especially during times of the year such as early summer when water is plentiful. So long as all out-of-stream diversion rights are being met and water diversions are not being measured, this practice goes unregulated by watermasters. When water becomes scarce, however, water users are legally entitled to ask their watermaster to regulate other users according to the priority scheme. This complaint system works well for out-of-stream diversions and other water uses for which owners are identifiable, as these owners will be motivated to immediately "call" for water when they are not receiving their share.

Instream rights, however, are held by the State of Oregon in trust for the people of Oregon. As trustee, the state is charged with complaining on behalf of public instream rights. In other words, the state must complain to itself when state watermasters are not enforcing instream rights. This is an unlikely prospect. On the other hand, a lessee of a water right for temporary instream transfer has the same legal standing as the lessor regarding management and enforcement of the instream right. Short-term instream leases therefore have an enforcement advantage over public instream rights. Lesees can complain on behalf of their rights, however, they may nonetheless be hampered by a lack of accurate streamflow information.

"Creative" Administrative Challenges to Instream Transfers

On top of the practical obstacles to instream protection, OWRD's administrative interpretations of Oregon's transfer laws make life difficult for instream applicants. Three administrative policies in particular have the potential to discourage future instream transfers. These include limiting total instream rights on a river to “estimated average natural flow,” refusing to create instream rights greater than state-created instream rights, and challenging the creation of permanent instream rights at the end of a lease period based on the “ready, willing, and able” language. These policies are not mandated by Oregon’s water law; rather they represent the choice of OWRD to scrutinize instream transfers and, more importantly, epitomize the divergent handling of instream and out-of-stream rights.

Estimated Average Natural Flow as a Ceiling for Instream Rights

OWRD’s rules limit the amount of an instream right to the “estimated average natural flow” (referred to by the misnomer “ENAF”) of the river. This number varies by month, and for most rivers in Oregon it decreases throughout the summer. For example, the hypothetical “River Why” in Eastern Oregon has an ENAF in June of 3 cubic feet per second (cfs), in July of 2 cfs, and in August of 1 cfs. If a landowner on the River Why had a 2.5 cfs irrigation right they wished to transfer instream, OWRD’s rules would limit the instream right in July and August of a given year to 2 and 1 cfs, respectively, regardless of how much water was actually available. In other words, during a wet year, if 4 cfs flowed instream through August, the instream right could nonetheless be limited by ENAF potentially allowing irrigators with junior rights to diminish flow instream below 2.5 cfs. OWRD allows for an exception where “flows that exceed [ENAF] are significant for the public uses applied for. An example of such an exception would be high flow events that allow for fish passage or migration over obstacles.” Two questions are raised by this hypothetical. First, how would a landowner have a water right greater than ENAF in the first place? Second, ENAF doesn’t seem to be much of a limitation. If the water is not normally available anyway, how does it matter that the instream right is limited?

The answer to the first question is simple. When many senior water rights were established, no one knew what the ENAFs for particular rivers were. Therefore, many rivers in Oregon are overappropriated the sum of the water rights is greater than the average flow of the river at any particular time. The second question, related to the first, can only be answered by comparing the treatment of the instream right with how the right would be treated if it remained an irrigation right. Out-of-stream uses are not limited by ENAF. On the hypothetical River Why, therefore, in a wet year with 4 cfs flowing through August, a 2.5 cfs out of stream right would be satisfied through August, assuming all senior rights were being met as well. In contrast, OWRD could allow irrigators with rights junior to the instream right, to withdraw water, reducing the instream right to ENAF 2 cfs in July and 1 cfs in August. This result is not fair; after all, when creating instream rights, the legislature clearly stated that instream rights should be on equal legal footing with "any other right for which a certificate has been issued."

OWRD's rationale for limiting instream fights to ENAF is that adding water instream above the average natural flow is not a beneficial instream use. This may be a valid legal argument it may seem like too much to ask to protect more water instream than average natural flow. However, this argument assumes no marginal benefits accrue above average flows. In reality, above average flows could result in above average fish survival or other related ecosystem benefits. Zero flexibility in the ENAF policy unfairly precludes these potential benefits and further tilts the field toward consumptive users. Out-of-stream users, even on over-appropriated rivers, are only regulated off according to priority. Imposing the ENAF limitation is an unnecessary burden on instream rights.

Another "Beneficial Use" Ceiling

In addition to the ENAF ceiling limiting new instream rights, OWRD may also limit the conversion of existing rights to instream rights in relation to the amount of state-created instream fights on the river. As noted earlier, since 1987, the departments of Fish and Wildlife, Environmental Quality, and State Parks and Recreation have been able to ask OWRD to create instream water rights to promote specific "public uses." Fish and Wildlife requests instream rights on behalf of the public for "conservation, maintenance and enhancement of aquatic and fish life, wildlife and fish and wildlife habitat. " The Department of Environmental Quality does so in the name of "pollution abatement," and Parks and Recreation in the name of "recreation and scenic attraction." In conjunction with these statutorily defined "public uses" OWRD will only approve instream rights that "serve a public use or uses" and will not allow the combined amount of instream fights for a reach to "exceed the amount needed to provide increased public benefits." In other words, once an instream right is granted on a fiver or through a reach for a purpose requested by Fish and Wildlife, Environmental Quality, or Parks and Recreation, the right becomes the maximum OWRD will protect instream on that fiver or reach.

In practice, the result of this policy is that instream transfer applications by private landowners with senior rights are allowed only up to the cap set by the state created instream fights. For example, if Fish and Wildlife applied for and was granted a 5 cfs right on the hypothetical River Why in 1988, the total amount of instream rights on the River would be capped at 5 cfs. A 2 cfs instream right created by temporary lease, with a senior date of 1909, would not augment Fish and Wildlffe's instream right, but instead be subsumed within it. The total amount of instream rights on the River Why would remain 5 cfs but 2 cfs of the rights would have a senior priority date of 1909 and 3 cfs would have a priority date of 1988. Another applicant who then sought to transfer an additional 4 cfs instream on the River Why would be limited to 3 cfs such that the combined instream rights do not exceed the 5 cfs specified by Fish and Wildlife.

Application of the policy operates differently when the instream right requested by a state agency exceeds the current amount of available unappropriated water. In this case, OWRD uses the instream amount requested as a "management objective." As landowners apply for instream rights on the river, those rights will accumulate toward the "objective" amount of instream water until the goal is met. Once the objective amount is reached, OWRD would cap instream rights on the river as outlined above.

Nothing in the Instream Water Rights Act requires a cap on the combined amount of instream rights. OWRD may be misinterpreting "beneficial use" in imposing this limitation. When creating instream rights, the legislature defined two important terms. First, "instream flow" is "the minimum quantity of water necessary to support the public use requested by an agency." Second, the legislature made clear that "public uses are beneficial uses." OWRD reads these provisions together to mean that instream rights, whether state-created or created by private-party purchase or lease, are only beneficial to the extent they support the minimum amount necessary for a public use. The legislature did not impose this limit, and a plain reading of the law does not require it. Nowhere does the Act refer to the combined amount of instream rights in a river or reach and limit that amount to the instream flow requested by state agencies. In fact, the language giving private parties the right to lease water rights for conversion instream is devoid of reference to public uses, instead declaring simply that "the use of the [leased] water right as an in-stream water right shall be considered a beneficial use." A plain reading of the statute, then, supports two separate types of instream rights: State agency-created rights limited to the minimum quantity necessary for public uses and rights created by private-party lease or purchase which are beneficial uses separate of their relation to public uses.

OWRD's handling of state agency instream rights makes the assumption that the agencies "hit the nail on the head" every time they request instream rights. Yet fish biologists, for example, disagree about the interaction between streamflow and fish habitat. As a result, an instream request for fish habitat could prove inadequate in the future as streamflow needs are better understood. The law also lacks any standards for the agencies to aspire to, except that, within their discretion, they "may request" instream rights to support public uses. Because the agency-requested rights can effectively become caps on allowable instream rights, an arbitrary request could have far-reaching consequences limiting in perpetuity the total amount of possible streamflow protection.

The purpose of this discussion is not to say OWRD's policy is illegal and should be overturned. Rather, the point is to shed light on some of the weaknesses of OWRD's policy choices and demonstrate again the hesitancy of the agency to fully embrace the idea of instream flows for their own sake. Oregon law, through OWRD's policies, allows out-of-stream diversions to completely dewater numerous rivers throughout the state. Common sense dictates that the opposite should also hold. Oregon law and OWRD's policies should allow for a river where all out-of-stream diversions are converted to instream rights resulting in zero diversions and year-round water flow.

A Creative Application of the Forfeiture Law

Oregon's limited forfeiture statute exempts from forfeiture any amount of water use less than the certificated amount of the appropriator is nonetheless "ready, willing, and able" to put the full certificated amount to use. In other words, if, at the time of a transfer, an appropriator no longer has an irrigation system capable of handling the full amount of their water right, they forfeit the amount of water they are no longer "able" to use. OWRD may use this language to limit landowners' ability to permanently transfer rights instream at the conclusion of a five year lease period. The limitation would go something like this: the watermaster would challenge the instream transfer alleging that, because the landowner's irrigation system sat unused for ten years, the landowner is no longer "ready, willing, and able" to put their entire right to use. Alternatively, the watermaster could challenge the transfer on the grounds that the water right had been forfeited before the instream lease period began. They would then determine how much the dilapidated irrigation system could handle and limit the instream transfer to that reduced amount.

This limitation is a creative application of the law, at best. At worst, it represents an expression of hostility toward conversion of existing uses to iustream uses. Instream uses are equal in the eyes of the law to other beneficial nses. Use of the entire right instream during the lease period does not begin the forfeiture clock the right is in "use," and only non-use leads to forfeiture. The "ready, willing, and able" language is only germane to water rights requiring physical diversions and should not be applied to instream rights, which are fundamentally different. Specifically, when the Oregon legislature created instream rights one of the founding tenets of Western water law had to be unearthed and rearranged. The requirement that a landowner physically divert water from the stream to have a valid water right could not coexist with instream rights fights which by their nature are not diverted out of the stream.

The dichotomy between the physical diversion required for consumptive uses and the lack of such a requirement for instream uses highlights why the "ready, willing, and able" language of the limited forfeiture law should not apply in instream transfer scenarios. Water is always ready, willing, and able to flow down its natural course and the legislature has declared that water flowing in its natural course is essentially being put to "use." In contrast, putting water to consumptive use requires controlling works such as dams or pumps to transport water out of the river. The limited forfeiture law was, in part, an attempt by the legislature to encourage conservation by allowing irrigators the flexibility to reduce water use for example by switching to lower water use crops without the threat of losing some of their water right. The flexibility came with some costs however. Foremost was the requirement that, to retain their entire water right, irrigators had to maintain a system capable of putting the whole right to use should they want to switch to a more consumptive use in the future. In contrast, if an irrigator decides to transfer their right instream permanently after a lease period during which they used their whole right instream, the "ready, willing, and able" language should not apply. Doing so takes the limited forfeiture law entirely out of context and is a misreading of Oregon law.

Formal Opposition to Instream Transfers

Oregon law provides a forum for water users to oppose proposed transfers and present evidence of potential injury from the proposed changes. Formal proceedings, called "contested case hearings," are held in accordance with Oregon's Administrative Procedures Act (APA). The formal procedure is not the only method by which instream transfers are opposed, however. Opposition voiced by local watermasters prior to initiation of formal transfer proceedings can stop an instream transfer as well. Both forms of opposition demonstrate the vigour with which OWRD undertakes injury analysis in the face of potential instream transfers. On the other hand, understanding how OWRD analyses injury in instream transfers will allow future applicants to come to the process better prepared and with applications better tailored to comply with the no injury rule.

The Water Trust's Contested Case Hearing

In 1998, the Oregon Water Trust applied to transfer 0.341 cfs of irrigation fights dating from 1863 and 1925 to instream use on behalf of a landowner in Union County, Oregon. Both rights authorised diversions from Little Creek, a tributary of Catherine Creek, which in turn flows into the Grande Ronde River. The 1863 right was for irrigation use and was used intermittently throughout the year. The Oregon Water Trust proposed protecting the water instream as a reach for 2.5 miles from the old POD on Little Creek to the confluence with Catherine Creek and changing the intermittent use of the 1863 right to continuous flow from June through September at a lower flow rate. A group of landowners and a local ditch improvement district (the "protestants") filed a protest of the proposed transfer. All of the protestants' rights are junior to the applicant's 1863 right, and with the exception of the improvement district, all of the protestants' water rights are upstream of the proposed instream reach. In addition, none of the protestants rely on return flow from the applicant's lands for their water rights. The protestants' opposition to the transfer was based on the fear of being regulated off during the irrigation season to protect the more senior instream right downstream of their diversions. In response to the protest, OWRD initiated a contested case hearing to decide the matter.

Contested case hearings, held under Oregon's APA, are quasi-judicial proceedings during which evidence is presented, including witness testimony and cross-examination. According to the APA the burden of proof in a contested hearing "rests upon the proponent of the fact or position." The applicant for a transfer therefore has the burden of proving that no injury will occur if the transfer is approved. However, because the purpose of a contested case hearing is to develop a full record upon which a decision on injury can be made, the burden of proof is "not so explicitly assigned." The primary concern in the water right context is that the hearing develop a record upon which OWRD can make an affirmative finding by a preponderance of the evidence that "the proposed change can be effected without injury.. "

The protestants presented two main arguments against the transfer: 1) the proposed transfer would result in more water being left instream than was historically used on the applicant's land, and 2) the instream right would change the historical pattern of use and result in reduced seasonal availability of water to downstream irrigators. The question before the hearing officer was whether the changes protestants feared constituted legal injury. As noted above, differentiating legal injury from allowable changes is the crux of the injury analysis. The Oregon Water Trust and OWRD contended the changes did not constitute legal injury because they were no different than the impacts protestants would suffer had the applicant exercised his water right to its hill "paper" extent. The hearing officer agreed with this interpretation and issued a "Proposed Final Order" stating "the transfer can be affected without legal injury to existing water rights."

A number of key "guideposts" for injury analysis in formal transfer scenarios can be gleaned from the Proposed Order and the testimony during the hearing of Thomas J. Paul, the Administrator of the Field Services Division at OWRD:

Regulation of junior rights is not legal injury : The protestants' major concern was that after creation of the instream right, they would be regulated off from use of their rights to a greater extent than in the past. Again, all of the users but the irrigation district were not only junior, but also upstream of the proposed instream right. Because of their junior priority dates, however, the protestants could have been regulated off at any time regardless of whether the applicant's right was an instream right or an irrigation right. Prior appropriation is based on a system that regulates in favour of senior rights at the expense of junior rights. Inherent in this system is the inevitable shut-off of junior diversions to allow water to pass to seniors. While this type of regulation may cause hardship, it does not represent legal injury. In the context of instream rights the Oregon legislature conferred upon them equal legal status with other types of appropriations. As such, if an instream right is senior, regulation in favour of the right to the detriment of juniors is not legal injury.

Water right transfers in oregon begin with the presumption that the entire "paper" right is transferable : The protestants also argued that the proposed regulation of the new instream right would result in protecting more water than was historically used by the applicant. Again, under Oregon law, this is not legal injury. Oregon's transfer statute is explicit that water subject to transfer includes the water represented by a "water right certificate." This language creates the initial presumption that transfers begin with the amount printed on the certificate. Oregon's law contains no reference to historical use that would limit transferable water to what was historically used by the appropriator. Further supporting this proposition, Oregon's limited forfeiture law exempts from forfeiture any amount of water less than the certificated amount of the appropriator is nonetheless "ready, willing, and able" to put the full certificated amount to use. In other words, using less than your "paper" right does not work a forfeiture ff you are otherwise capable of diverting the entire amount. OWRD interprets these statutes as mandating that, "in evaluating transfers, the Department begins with the full face value of the water right and then examines the proposal to determine whether it would cause injury." Therefore, unless OWRD determines injury will occur, it must approve a transfer of the full "paper" water right.

A water right holder is not entitled to the undiverted portion of another right holder's water : Based on Oregon's limited forfeiture law, downstream users such as the improvement district in the Little Creek dispute who are accustomed to receiving an undiverted portion of an upstream user's water right are not legally entitled to continuance of that amount of water. Oregon law allows a change from a historic use that consumed a fraction of a water right to a use that consumes the entire water right and leaves no excess water flowing to downstream users as long as no legal forfeiture has occurred. This includes transfer from a consumptive use to instream protection. The undiverted portion of an upstream users' right should not be confused with established return flow from an upstream user to which downstream users are legally entitled. In the Little Creek dispute, none of the parties could identify return flow from the applicant's land and the dispute was over undiverted water. Return flow is limited to water returning to the source stream after use and is not the same as undiverted water. Once again, what instream proponents seek is equal treatment with consumptive uses. The limited forfeiture law does not discriminate between consumptive and instream uses and the formal protest process should reflect this.

Changing the historic "shape" of a water right is not injury if the change is within the extent allowable under the right : An important part of the Oregon Water Trust's application for the Little Creek instream right involved changing the use from intermittent use to continuous flow so that the water would be protected, i.e. in "use" for a continuous but shorter time period. Changes in the parameters of a water right such as timing of use are often referred to as "shaping." Such shaping of a water right is not injurious if the new "shape" "could have been affected by the exercise of the water right to the fullest extent allowed by law." Water rights certificates specify the rate at which an appropriator may divert water, the full extent of the monthly or seasonal duty (expressed as the total amount of water that may be applied to each acre of appurtenant land each month or for the season if one is specified), and the season of use of the water right. Enlargement, a species of legal injury, occurs if the reshaped water right exceeds any of these parameters. However, changing the historic timing of use as the Oregon Water Trust proposed for the Little Creek right, is not injury because the applicant would have been within his "paper" right to make the change regardless of the transfer. In other words, the applicant was not proposing to use more water or use the same amount of water for a longer period of time. Instead, the new "shape" of the water use would be a lower but continuous flow for a concentrated period of time rather than intermittent use throughout the year.

Continued "subirrigation" from adjacent lands after an instream transfer does not constitute enlargement : In the Little Creek dispute, the protestants claimed that the applicant's land Could not be prevented from receiving water from Little Creek even after transferring the entire right instream because his land would continue to be subirrigated from neighbouring lands and because a wetland fed by Little Creek could not be dried up and would continue to seep water onto the applicant's land.

Under Oregon's administrative rules, a water right will be considered enlarged if under a change in place of use, the original place of use cannot be prevented from receiving water from the same source. This rule prevents users from transferring a water right to a new place of use in such a way that the old place of use continues to benefit from subirrigation by water applied on the new place of use. OWRD interprets this rule to mean that if the original place of use continues to benefit from the transferred water right, enlargement occurs. But under OWRD's interpretation of its own rules, the "same source" language refers to "water diverted from the water right being transferred,"the specific water that was previously being diverted and irrigating the transferor's land and excludes subiiTigation from water rights used on adjacent land or from springs or wetlands not included in the water right being transferred. Therefore, because the applicant's land in the Little Creek dispute would receive only limited subirrigation from adjacent lands and springs—natural phenomena that could not reasonably be prevented the hearing panel found no enlargement would occur.

Efficiency of a water user's conveyance system is not an injury consideration : Finally, the protestants argued that once the applicant stopped diverting his irrigation water, those users who shared the applicant's irrigation ditch would not receive their entitled water because the applicant's diversion was necessary to help "carry" their water down the ditch. On Little Creek, as in many other places, water users often share one POD and one conveyance system such as a ditch, canal, or pipe, with each landowner taking his water from the system as it passes his land.

Because of inefficiencies in conveyance systems such as leaky pipes, surface evaporation from open ditches, or seepage losses out of unlined ditches, together referred to as "carriage losses," transporting a given amount of water from the source to the place of use can be difficult. Often, diversion of more than the sum of the water rights the system serves is required to provide sufficient "head" to transport the water to the respective places of use. Maintenance of a diversion and conveyance system capable of diverting a user's full water right is the responsibility of the user. More importantly, water rights certificates specify the rate of diversion at a specific point on the source stream. Because water right certificates specify the rate at the point of diversion and not at the place of use, watermasters are charged with ensuring the specified amount of water is available at the POD, but are not charged with ensuring that amount of water makes it to the place of use. In other words, no injury occurs because an inefficient conveyance system like the ditch in the Little Creek dispute is made more inefficient when one user ceases diverting their share through the ditch.

The foregoing discussion is an attempt to shed light on many of the issues encountered in formal contests of instream transfers. The discussion is by no means exhaustive. As evident in the next section, new challenges lie around every corner. However, keeping these "guideposts" in mind at the outset can help an instream transfer applicant anticipate and mold their application around potential difficulties. The next section further illuminates the vigour with which OWRD applies formal injury analysis to instream transfers.

High Water: Watermaster Opposition

The Walla Walla River originates in the Umatilla National Forest in far northeast Oregon and flows twenty-eight miles before crossing the border into Washington where it spills into the Columbia River. A landowner with 100 acres adjacent to the Walla Walla River approached the Oregon Water Trust desiring to transfer water instream from nineteen of the acres. Under a water right certificate issued in 1895, the applicant irrigates land as pasture for livestock. The water is delivered to the land through a one and a half mile long, unlined, earthen ditch. The rate of diversion of Walla Walla River water associated with the nineteen acres is 0.7125 cfs measured at the POD of the ditch.

With the Trust's help, the landowner submitted an application to OWRD to effectuate the instream transfer. In consultation with the local watermaster, OWRD sought to reduce the amount available for transfer instream to account for loss of return flow. No formal protests were filed, and the proposed reduction occurred in the context of OWRD's initial review of the transfer. The watermaster based his calculations on the Cuenca tables' model of crop water requirements. He calculated the consumptive use of nineteen acres of the applicant's pasture based only on the crop use from transpiration and set that as the amount available for transfer, calling the entire difference return flow. This calculation assumes all water not directly used by the applicant's crops is return flow available to downstream irrigators. During July, when the pasture crop is using the most water, OWRD has indicated they would limit the instream right to roughly twenty-five percent of the total certificated amount or 0.179 cfs. After July, the instream right would decrease further as crop transpiration decreases with the cooler fall weather.

The applicant and the Trust hoped to put more than 0.179 cfs back into the river during the driest months of the summer. The Trust therefore hired consultants to evaluate return flow from the pasture back to the Walla Walla River. Based on the consultant's calculations, some return flow from the pasture does find its way back to the river, but the time it takes for the water to do so ranges from twenty to fifty days depending on where the water is applied. For their calculations the consultants assumed an irrigation season beginning in March and ending October 1. October 1 is a common ending point for irrigation seasons in Oregon. Under this assumption, water applied to acres furthest from the river during the later months of the summer does not return to the river before October 1, after which, the Trust assumed, it would no longer be useful to downstream irrigators. As a practical result, the Trust felt this water should not be subtracted as return flow and should therefore be available for transfer instream.

Specifically, the Trust's consultants calculated that water applied 400 feet from the river takes twenty days to return to the river through underground channels, water applied 700 feet distant takes thirty-five days, and water applied 1000 feet away takes fifty days to reach the river. The amount of the total diversion applied 1000 feet from the Walla Walla River is 0.3652 cfs. The consultants then calculated the latest dates of water application at the various locations that would allow return flow from these lands to reach the Walla Walla before October 1. For land 400 feet from the river, water applied after September 11 will not reach the river before October 1. Water applied 1000 feet distant after August 12 will not return to the river before October 1. Using these dates combined with the amount of water applied at each distance interval, the consultants calculated that, from August 12 to October 1, 0.39 cfs is effectively "consumed" because it does not return to the river before the assumed end of irrigation season. The Trust felt this number represented an amount of water that could be protected instream beginning on August 12 and avoid injury to downstream users. Before August 12, the consultants recommended using OWRD's consumptive use numbers from the Cuenca tables to set the amount of instream protection. The Trust and the applicant did not object to these lower numbers during the early summer months because flows are higher in the Walla Walla during these times and instream protection is not as vital.

The Trust and the applicant were confident they had found a solution balancing protection for downstream users with intelligent, site-specific water management for instream benefit. OWRD did not agree even after reviewing the Trust's hydrologic report. Instead, OWRD raised a new challenge to the transfer from a different angle.

OWRD's new argument centred on the lack of a limited irrigation season on the Walla Walla River. The outer limits of irrigation seasons in Oregon are set by court decree. The irrigation season for the Walla Walla River is unrestricted. According to the decree:

Unlike most irrigated sections, the water users in the Walla Walla Valley apply the water to their lands throughout the year. Such use tends to saturate the ground and build up the water table so as to retard and reservoir [sic] the water, making it available for use later in the season. This practice appears to be beneficial and necessary for the production of good crops.

The Trust's proposed solution incorporated a limited irrigation season under the assumption that downstream irrigators would cease diverting some or all of their right after October 1 when wet weather returns. Based on the decree's unrestricted irrigation season, however, OWRD claims the water, whenever R returns to the Walla Walla even of it returns after October 1 should be available for downstream users. In other words, OWRD does not agree with the Trust's assumption that return flow is not useful to downstream irrigators if it returns after October 1. In light of this disagreement, the question should be, "did downstream irrigators in the past, in fact, rely on return flows from the applicant's lands after October 1?" ff the answer is affirmative, OWRD is reasonable in continuing to limit the instream right after August 12. If, however, downstream irrigators have never depended on the contribution of return flow from the applicant's land after October 1, continued OWRD opposition would be another example of the double standard on instream transfers. Finding a definitive answer requires a factual determination of historical irrigation practices in the area. However, the Trust's assumption that irrigation ends on October 1 is likely supportable by October, temperatures are moderating, rains are returning, and downstream irrigators are less likely to rely on the small contribution in return flow from the applicant's land. No definitive answer is necessary, however, for the discussion to provide important guidance. Four more "guideposts" for instream transfers arise from the scenario.

The standard for proving return flow requires more than mere speculation : As the Walla Walla scenario demonstrates, the presence of return flow depends on a number of site-specific hydrologic and topographic variables and cannot be theoretically assumed or taken for granted. Site-specific analysis is necessary to determine the existence, amount, and timing of return flow. Basin-wide observations are insufficient and the person opposing the transfer on the basis of return flow must offer more than "mere speculation" about the flows' existence. Therefore, in the face of opposition from OWRD, the applicant seeking an instream transfer should ask OWRD to provide more than speculative evidence more than the local watermaster's "gut" feeling that there is return flow.

Timing of return flow must be taken into account : When water is removed from a source either permanently or for long enough that it returns to the source only after irrigation ceases for the year, that amount of water should be counted as consumptive use and should not be subtracted from a proposed instream transfer as presumed return flow available to other users. Watermasters are often more conservative than this, using the Cuenca tables to determine crop [E_T] of specific crops in a transfer proceeding and using this number alone to determine how much water is available for transfer. The flaw in this approach is that it assumes all water not consumed by plants returns to the source, and does so in time for downstream users to divert and use the water during irrigation season. Crop ET should be a starting place in determining injury based on the presence of return flow, but relying on this number alone oversimplifies the analysis. To the extent that return flow will not return to the source before the end of irrigation season, the entire water right should be protectable instream.

Instream applicants should pay close attention to irrigation season limitations : Analysing the timing of return flows requires that the applicant know whether an official irrigation season is set by decree or included in the water right itself, or alternatively, what the prevailing local practices dictate. In the Walla Walla transfer, the Trust ran headlong into the problems caused by a year round irrigation season. If the Walla Walla decree, like many other decrees in Oregon, ended the irrigation season on October 1, the Trust's model would have met less resistance from OWRD. Knowing the irrigation season is also important for applicants seeking to transfer a right instream and change the "shape" of the right because the shape will be restricted by the irrigation season if it is limited.

Return flow entering the river at a point above a downstream junior's pod should be subtracted to the extent of potential injury : With all that has been said about when return flow should not be subtracted from an instream transfer, it would be irresponsible to ignore situations where return flow should be subtracted to protect other river users. Proven return flow—flows demonstrated by affirmative evidence should be subtracted from the amount of water available for transfer to avoid injury to downstream appropriators. This is appropriate for transfers to either instream or different consumptive uses or places of use. OWRD's policy "works backward from the full face value of a water fight, presuming all other transfer application criteria are met, limiting the proposed transfer to the extent that injury is found." For instream transfers in particular this means reducing the amount that will be protected instream by the amount of proven return flow legally and actually relied upon by other water users. However, this process need not be static; dynamic compromises are possible when parties are willing to work together. For instance, instream rights may be separated into distinct reaches defined by where return flow enters the river. If the exact location of the return flow is known, the full amount of the water right may be protected instream from the old POD down to this point, and the instream right can diminish thereafter. On the other hand, if the exact point of return for return flow is unknown, it may be appropriate to reduce the instream right at the old POD to account for uncertainty. However this is done, instream transfer and consumptive use transfers should be treated the same.

Achieving Streamflow Protection by Modelling Success

The final section of this Comment will explore the future of streamflow protection in Oregon. Perhaps the greatest step Oregon could take toward more effective management of water rights both in and out-of-stream is implementation and strict enforcement of a comprehensive metering and reporting requirement. The impact of such a law would likely be immediate rivers dewatered by wasteful, unregulated consumptive uses would have the chance to flow again and junior instream and out-of-stream rights that have gone unsatisfied for decades might be fulfilled. With a clearer picture of actual water use and availability, resource management decisions could be more responsive and accurate. That said, large-scale changes like comprehensive metering take time and streamflow protection cannot wait. Creative solutions within the existing legal and regulatory context are necessary.

The evolution of the water trust concept, employing cooperative, market-based, approaches to arrive at these solutions, is a primary innovation. Two examples in the following discussion illustrate the creativity necessary to increase streamflow protections under the current regime. In the John Day Basin in central-eastern Oregon, the Oregon Water Trust recently finalised a deal on the Austin Ranch to shorten the irrigation season for the use of several water rights, thereby leaving water instream during key parts of summer. This transaction exemplifies creative application of the transfer laws respecting existing rights while increasing streamflow protection. Farther north and east from the John Day Basin in the Lostine River Valley, the Oregon Water Trust brokered another unique deal with a large number of landowners who agreed to limit their late-summer diversions to maintain a specified amount of water in the Lostine River. This deal demonstrates the potential for purely private agreements to achieve instream protection outside the boundaries of the law.

The next sections discuss three successful examples of the above-mentioned strategies. First, Washington State's measurement program is analysed. Next, the two Water Trust deals are examined in detail.

A Measurement Model

In 1993, Washington enacted a requirement of "metering of diversions or measurement by other approved methods" as a requirement for all new surface water rights and for some existing rights as well. The law requires measurement of new surface water permits, new and existing surface water rights in waters containing depressed or critical salmonid stocks, new and existing groundwater rights where withdrawal could affect depressed or critical salmonid stocks, and existing surface water fights where the diversion exceeds 1 cfs. In contrast, Oregon has no comprehensive surface measurement requirement. Through permit conditions and other sporadic regulations, Oregon measures only 8% of combined surface and groundwater rights accounting for only 46% of the water subject to diversion. Much of this measurement is of municipal water use and groundwater wells. OWRD has also inventoried "significant" diversions, but has yet to impose measurement requirements.

The ineffectiveness of Oregon's measurement program stems from the types of water uses measured. Small, salmon-supporting creeks in rural parts of the state, which is a majority of the streams that groups like the Oregon Water Trust work on, have few municipal water withdrawals. Instead, these streams are dewatered by small irrigation diversions, often diverting more than their certificated right. Without a comprehensive metering requirement, irrigators are left to divert as much water as their ditch can handle. Unless another user calls for regulation, this unlawful practice often goes unnoticed. Non-enforcement against unlawful diversion is the most significant symptom of Oregon's measurement failure and instream rights will bear the brunt of the impact. Relying on serf-policing by irrigators lacking accurate information and with little incentive to minimise diversion and use results in instream rights throughout the state not being met despite water availability.

By 2002, Washington had issued administrative orders for 903 water rights representing 80% of total estimated water diversions in its fish critical basins. Perhaps more importantly, Washington provided more than $3,000,000 in financial assistance to help defray the costs of installing metering equipment. Both the number of administrative orders, and the financial commitment demonstrate that Washington has a significant head start over Oregon in instituting meaningful water measurement. Oregon's current scattered measurement requirements lack focus, comprehensiveness, and financial backing. Oregon would benefit from a more targeted measurement approach modelled on Washington's laws. Specifically, OWRD should focus on small, fish producing streams where incremental increases in instream water would have large impacts.

Changing Use Without Risk of Injury

While many instream transfers are limited by injury considerations to small amounts of water, the Oregon Water Trust recently completed a deal on the Middle Fork of the John Day River and two important tributaries to protect as much as 10 cfs instream during the driest months of the summer.

Beginning July 20 of every year, Austin Ranch, with a senior water right, stop diverting water from the Middle Fork and from two mountain tributaries. Shortening their irrigation season may require Austin Ranch to reduce the number of livestock they raise by approximately 20%. The Trust has compensated them monetarily for this change. The agreement is a simple and permanent change in the season of use of the water rights. To effectuate the change, Oregon Water Trust and the landowner submitted an affidavit to OWRD to "abandon" use of the water right during the late summer months. OWRD will not need to regulate other water users to enforce the 10 cfs instream, so injury considerations did not affect the change.

The added flow late in the summer Hill ensure that downstream irrigation withdrawals do not dewater the river as long as the downstream irrigators stay within their legal limits. The Oregon Water Trust and the owners of the Austin Ranch are confident the deal Hill result in "real" streamflow. No new appropriations are allowed in the John Day Basin so the water that was previously diverted by the ranch is not subject to new appropriation downstream. Even if the water were subject to appropriation, the adjacent downstream water right holders are conservation entities such as the Nature Conservancy and the Warm Springs Tribe who are not interested in out-of-stream uses. The next out-of-stream user is twenty miles downstream. Most importantly, the project has qualitative benefits on top of any increase in water quantity. Increased flows in the two mountain tributaries Hill improve fish habitat by adding much-needed cold water to entice migrating salmon. These qualitative benefits may be felt up to seventy miles downstream.

Private, Market Solutions on the Lostine River

The Oregon Water Trust's work on the Lostine River in northeastern Oregon is unique, involving a private agreement not requiring OWRD approval and an unheard of number of participant-landowners. Seeking to avoid federal regulatory intervention on behalf of listed anadromous fish species, irrigators along the Lostine River met with the Oregon Water Trust to discuss options for increasing summer streamflows. The result of these meetings was a deal involving five ditch companies and more than one hundred landowners.

Rather than pursuing short-term, instream leases, each involving a potentially complicated transfer analysis, the parties agreed to enter a contract requiring no OWRD approval. With the help of local fish biologists and Indian tribes, the parties set a goal of keeping 15 cfs instream during the late summer. Accomplishing the desired streamflow involved careful monitoring and day-to-day adjustment of the amount of water diverted out-of-stream. OWRD participated by installing a stream-flow gauge above the irrigators' diversions and paying an independent hydrologist to monitor flows twice daily. The hydrologist relayed his flow readings to the heads of the five ditch companies encompassing all of the irrigators. Based on these readings, the irrigators controlled their withdrawals always trying to leave 15 cfs instream. In return, the Oregon Water Trust compensated the irrigators for each day during the contract period that the goal was met.

The Lostine deal is remarkable in two ways. First, using private contracts outside OWRD's leasing program was previously unheard of. The high participation by irrigators demonstrates the power of using individualised market incentives to drive water apportionment decisions. Protecting habitat has market value to landowners who might otherwise face sanctions. At the same time, the cooperative solution respected the landowners' way of life leaving the ultimate decision of whether to meet the daily goal in the hands of the irrigators. Second, the Lostine deal highlights the potential of comprehensive metering and reporting to promote effective water distribution. Real-time knowledge of streamflow conditions was essential for the landowners to meet the daily goal. The task is similar throughout the state. The constant flux of water availability necessitates accurate measurement and, in turn, accurate measurement ensures impartial and effective distribution.

Several drawbacks take away from the Lostine project's success. The deal was limited to one summer. Absent another agreement, the Lostine River is not protected for the future. The arrangement was also expensive. The Oregon Water Trust paid more than $100,000 for one summer of instream flow protection. This level of financing is not sustainable on a long-term basis. Finally, many irrigators were unsatisfied with how the water was apportioned. Some felt cheated when flows in the Lostine exceeded 15 cfs feeling that water was being left instream that should have been used for irrigation. Of the five ditch companies participating in the deal, only one was equipped with a modern diversion capable of accurate metering. If all the ditch companies had similar equipment, the resulting accuracy of distribution would obviate the need to overshoot the daily goal and would result in more water availability for irrigators.

Despite these drawbacks, the Lostine River deal, as well as the Austin Ranch deal, are models for achieving streamflow protection by working "outside the box." Any solution that balances instream uses with respect for existing consumptive uses involves trade-offs on both sides. Both of the Trust's deals show that the two sides, irrigators and streamflow protection advocates, are willing to make these sacrifices in the name of mutual benefit.

Recommendations based on the forgoing discussion are two-tiered. On the first-tier are recommendations to change the way instream rights are administered. These changes require institutional and political changes changes that take time and steady pressure from advocates. The second-tier of recommendations are modelled after successes within the existing regime. The successes on the John Day and Lostine Rivers exemplify the future of streamflow protection in Oregon and the West. Both solutions avoid injury to existing uses without necessitating OWRD scrutiny or implicating OWRD's administrative policy hurdles.

First-Tier Recommendations

Enact a Comprehensive Measurement Requirement : Washington's model and the success of measurement on the Lostine River underline the need for accurate measurement in Oregon. Improving the availability and accuracy of information about water withdrawals would have an immediate effect on the distribution of water. Unlawful diversions and waste would decrease and instream and consumptive water uses both would experience higher rates of satisfaction. The common criticism of this proposition is that it will be too costly either water right holders will be forced to bear the burden or the state will have to take money away from other important initiatives. However, spending money on measurement makes sense as a public expenditure because it avoids unseen costs associated with unlawful water use and provides valuable benefits. Before implementing its measurement requirement, Washington examined several areas within the state where intense regulation, including measurement, was already being implemented and found that fishery benefits alone could amount to upwards of $18.1 million in a single basin. Other benefits may be less quantifiable but no less important, such as increased water availability during future droughts.

Treat Instream Transfers Equally with Consumptive Uses in Transfer Proceedings : This Comment outlines numerous examples of disparate treatment of instream transfers despite the mandate that instream rights are legally equal. Several specific changes would make huge strides towards leveling the playing field: 1) Molding injury analysis to better reflect site-specific hydrology and patterns of use by codifying the standards and burdens of proof OWRD must meet to prove injury or enlargement; 2) doing away with caps on total instream rights such as the ENAF policy which discourage instream protection and are not applied to other consumptive uses and; 3) limiting application of the limited forfeiture law's "ready, willing, and able" language to consumptive uses rather than instream uses. The important thing to recognise about each of the policies these changes address is that they represent OWRD's choices. The law does not compel the strict application of the ENAF policy, nor does it compel application of the "ready, willing, and able" language to instream rights. Our water use priorities are slowly shifting away from total consumptive use and OWRD should embrace this shift and lead the way in encouraging it rather than shying away.

Second-Tier Recommendations

Strive for Creative Solutions Within the Existing Legal and Regulatory Regime : Asking for large-scale policy change is a long-term solution. In the short-term, success requires familiarity with the spectrum of potential challenges to instream deals and the foresight to maneuver through them. Where possible, projects should be designed that do not implicate the "no injury" rule. Projects that effectively put water instream without creating official "iustream rights," like the Oregon Water Trust's change in season of use on the Austin Ranch, avoid many complications. Before applying for an instream right, practitioners should ask if any changes can be made to the existing use to achieve the desired streamflow without implicating injury analysis. Possibilities include changing the season of use, split-season leasing, and outright cancellation of rights in basins where new appropriations are barred.

Look for Private Solutions : The Oregon Water Trust's Lostine River deal highlights a new avenue for streamflow enhancement. For emergency situations or other short-term water requirements, private contracts provide a powerful and flexible tool. The range of potential agreements is vast. Examples include compensating irrigators for serf-policing and maintaining instream goals, compensation for installing metering equipment and reporting use, or rotational irrigation agreements where each of several irrigators agrees to forgo water on a seasonal schedule. The key to finding these solutions is working with irrigators themselves to pinpoint where their interests and the interests of those advocating streamflow enhancements intersect.

Advocates for streamflow protection must recognise the sea of change instream rights are bringing upon Western water law. The disconnect between using a system honed to distribute water out of rivers, to leave water in rivers for ecological benefit, is obvious from the preceding discussion of the confluence of theory and practice. Each new instream transfer calls upon the parties and OWRD to stretch the limits of existing laws to accommodate new values while trying to protect old ones. In light of this, OWRD should be highly commended for working with instream transfer applicants and groups like the Oregon Water Trust. Oregon's status as the first state to create instream water rights gives it a valuable head start in the movement to shift away from total consumptive use to a more sustainable balance. As a result, the amount of water left instream in Oregon is increasing each year. However, without continued innovation in the regulation and administration of water both instream and out, the risk of reaching a plateau is real. Hopefully, the lessons learned at the confluence of theory and practice are a starting place for these changes.

Chapter 4

Snowshed Contributions to the Nooksack River watershed

In Whatcom County, Washington, as elsewhere in the western United States, demands on water resources are increasing. Most water-policy research concentrates on allocation and distribution, attending much less to the possible dynamics of the source regions: high-elevation snowsheds. Persons and organisations interested in agricultural, fishery, hydropower, municipal, and recreational water uses show keen interest in every drop that flows down the rivers and streams of the western United States. Indeed, in many cases, such as the Nooksack River Basin, in the northwestern corner of the North Cascades Range, Washington, rivers are already legally overallocated, and demand exceeds supply. More important, research suggests that snowfalls and water supplies are declining in the western United States and that they will be significantly influenced by climate changes through this century.

The research determines the contribution of high-elevation snowshed meltwater to the Nooksack River watershed, a contribution previously estimated to be very large during the dry summer. Knowing the proportional contribution of glacier water is important, particularly because the glaciers are shrinking rapidly in the region. Several studies in the western United States speak to diminishing glacier size, the result of several years of negative mass balance. Any moisture shortage noted over glacier surfaces must also occur over nonglacierised portions of high-elevation snowsheds; links among glacier mass balance, seasonal snowpacks, water supply, and climate changes are therefore critical. If glacial recession continues at the present rate, the glaciers and their contributions to summer meltwater may be permanently lost. In that case, summer streamflow s will decrease, and water-management policies within the watershed will be drastically affected.

The Nooksack Basin displays significant variability, including a glacierised alpine region, heavily forested mountainous uplands, and a highly developed lowland. The important role of the river system is increasingly evident as population growth, well-established agriculture and industry, a salmon population listed as a threatened species, and expanding water-based recreation and tourism stress a fully allocated surface water supply. Determining the limits of the basin's sustainable water supply requires an understanding of the complex watershed hydroclimate. Runoff displays significant spatial and temporal variation in the Nooksack Basin. Temporal heterogeneity arises from the intra-annual, interannual, and secular changes in temperature, precipitation, and other climatic factors.

Given the relatively small area of the watershed, the temporal variations generally occur in phase across the basin. Spatial heterogeneity results from the climatic, topographic, biotic, land-use, and pedologic variability within the basin. Dividing the watershed into subunits maximises the landscape homogeneity within each area and provides a framework for analysing runoff variations related to hydroclimatic differences between glacierised and unglacierised subbasins. My study estimates the contribution of high-elevation snowshed meltwater to the Nooksack watershed, in the belief that knowledge of hydroclimatic variability will contribute to designing land- and water-resource-management programs that address increasing competition among users.

The Nooksack Basin

The Nooksack Basin offers an opportunity to test the hypothesis that glaciers and adjacent snowsheds in the North Cascades contribute up to 25 percent of the summer streamflow. The Nooksack River drains an area of approximately 2,015 square kilometres. The drainage basin is dominated by two physiographic provinces: the Puget Lowland and the western slope of the Cascades. The low-relief Puget Lowland gently rises from sea level to about 300 metres at the foothills of the Cascades.

The western slope of the Cascades rises from the Puget Lowland to a series of ridges 1,000-2,000 metres in elevation, with the glacierised summits of Mount Shuksan (2,782 metres) and Mount Baker (3,284 metres) along the eastern drainage divide. Topography is of high relief, due to Pleistocene glacial erosion.

Figure : The trailhead to take pictures of Mount Shuksan and the Nooksack River.

Forests dominated by fir, cedar, and hemlock covered nearly the entire Nooksack watershed prior to European settlement, but they have largely been cleared for agriculture and urban development. The lower elevations of the upper basin have been logged since the 1930S. These physiographic provinces provide two different units of land use. The Puget Lowland is drained by the lower Nooksack River, west of the town of Deming, Washington. They have been developed into intensive agriculture (irrigated field crops and dairying), and their human population of about 47,000 utilised 183.2 million liters per day of surface water from the Nooksack watershed in 1995. The upper Nooksack Basin is strikingly rural, predominantly utilised for timber harvest, wildlife habitat, and outdoor recreation and with a human population of less than 1,000. The upper basin is largely federally managed land (47 percent) at higher elevations, with private (43 percent) and state (10 percent) lands at elevations mostly below 1,000 metres. The upper basin provides an average 92 percent of the basin runoff, whereas more than 99 percent of the water use by humans takes place in the lower basin.

Except for a diversionary weir for the city of Bellingham's public water supply on the Middle Fork, the upper basins contain no significant structures. The hydroclimate of the upper Nooksack Basin is strongly influenced by a winter precipitation maximum and a steep topographic gradient from west to east that accounts for relatively abrupt changes in the quantity and form of precipitation from rain to snow. The cold, very snowy environment of the high elevations contrasts markedly with the mild, rainy climate of the lower elevations to the west. Mean monthly precipitation and temperature from weather stations at Clearbrook and Glacier illustrate these contrasting energy and moisture conditions in the Puget Lowland and the Cascades. The linear distance between these two weather stations is 31 kilometres, but Glacier is 266 metres higher in elevation than Clearbrook. Glacier's average annual temperature is 1.2°C cooler than that of Clearbrook; and it receives 52.4 centimetres more precipitation, mostly in the form of snow, than does Clearbrook.

Moisture-laden Pacific air masses generally follow a northeasterly storm track over the Puget Lowland and into the Cascades. The jet stream directs cold front after cold front through the area during the winter. Slowed by the Cascades, these storms produce low-intensity, long-duration precipitation through orographic uplift. Snow levels vary throughout the winter and from year to year, but elevations above 2,000 metres commonly accumulate 8-10 metres of snow. Occasional warm southwesterly winds elevate ambient temperatures during winter storms, leading to rapid melting of snow that sharply increases runoff (rain-on-snow events). Summers are dominated by a ridge of high pressure over the Pacific, highly stable air, and low precipitation totals.

The watershed response in transforming precipitation into runoff is determined by the interaction of physiographic and hydroclimatic basin features. Two subunits were selected from within the watershed to provide a framework for analysing runoff heterogeneity related to hydroclimatic differences between glacierised and unglacierised subbasins. Subdivision of the watershed into spatial domains expected to act independently based on the watershed's physical characteristics is recommended to improve the estimate's ability to represent runoff processes.

The downstream limit of the glacierised subbasin in the upper North Fork Basin is defined by the stream gauge near the town of Glacier, at an elevation of 380 metres. The 271.9 square kilometre basin has a mean elevation of 1,311 metres, with 16.6 square kilometres (approximately 6.1 percent of its area) of glacier cover. The unglacierised basin in the upper South Fork Basin is nearly the same size (266.8 square kilometres) but much lower in mean elevation (914 metres) and relief. The highest peak in the subbasin (1,800 metres) does not support a glacier. The downstream limit of the unglacierised subbasin is defined by the stream gauge near Wickersham, at an elevation of 117 metres. The entire upper basin is represented by a stream gauge at Deming, downstream from the confluence of the subbasins and 66 kilometres above the river's mouth, at an elevation of 62 metres.

Daily discharge data from these three gauges were obtained from the U.S. Geological Survey Water Resources Division. The electronically obtained data are officially listed as unverified, but no anomalous flows are apparent. All data arc organised by water year (1 October-30 September) and are designated by the calendar year in which they end. Accordingly, water year 1965 began on 1 October 1964 and ended on 30 September 1965. Data from the period between 1 October 1937 and 30 September 1999 were selected for analysis based on the completeness of each record. The Deming station is missing data between 1 October 1957 and 30 September 1964. The unglacierised station is missing winter data between 1 October 1977 and 30 September 1995 but has complete data during the spring and summer months.

Descriptive statistics were calculated on daily discharge data and graphed to visualise seasonal fluctuations in each basin. Because the discharge means and standard deviations are temporally and spatially variable in magnitude, the data were standardised by calculating the coefficient of variation (the standard deviation divided by the mean). Differences in daily discharge values were calculated to show which sub-basin produced more discharge by subtracting the unglacierised basin discharge from the glacierised basin discharge. The two basins are roughly the same size, so differences in discharge represent heterogeneity of water inputs between them. Positive discharge values indicate that more water is produced in the glacierised basin; negative discharge values indicate that more water is produced in the unglacierised basin.

To calculate the percentage of the flow contributed to the lower Nooksack Basin by high-elevation snowshed meltwater, the difference in upper-basin discharges between the two subbasins was compared with the discharge of the total upperbasin contribution using the following equation: %M = (G-U)ID * 100%, where %M is the percentage of glacier and/or snow meltwater contributed by the glacierised subbasin; G-U is the difference in daily discharge between the glacierised (G) and unglacierised (U) subbasins; and D is the daily discharge of the Nooksack River downstream from all surface water contributions from Cascades, measured at Deming. Positive values of %M represent the percentage of snow and/or glacier meltwater contributed to the lower Nooksack River by the glacierised basir. Dividing the difference in discharge between the subbasins by the entire basin discharge, this estimate assumes that the variability in percentage contributions by all sources in the upper basin are the same, so that any variability in this value represents changing water contributions by the glacierised basin. Because the analysed basin contains a fraction of the high-elevation areas of the watershed, these estimates are minimum values for the percentage contributions from high-elevation snow-sheds in the Nooksack Basin.

Seasonal Runoff Regimes

The two subbasins exhibit annual discharge patterns that have seasonal similarities, yet important differences exist in flow characteristics within each season. Four seasonal flow regimes were identified in each basin: an autumn rainfall season, a winter snowfall season, a spring snowmelt season, and a summer low-flow (high-elevation snowshed melt) season. The homogeneity of seasonality between basins is indicative of the driving force of synoptic atmospheric conditions. The difference in discharge within a runoff season is due to the hydroclimatic heterogeneity between the two basins.

The autumn rainfall season follows the low-precipitation-low-discharge late-summer season. Typically, discharge begins to rise in October from rain produced by midlatitude cyclonic storms. This season is typified by runoff peaks on the hydrograph resulting from the passage of each frontal system.

The mean flow for each subbasin is between 11 and 17 cubic metres per second and gradually rises to a peak in late November as the polar front migrates into the region. This flow season is characterised by a high degree of variability in each basin, as indicated by the finding that the coefficient of variation nears or exceeds 1.0. In addition, the coefficient of variation shows considerable day-to-day variation, thanks to the chaotic timing of precipitation and melts events. Runoff peak flows can be larger in either basin, depending on the exact path of the storm, but the unglacierised basin averages a higher discharge for the season. The unglacierised basin shows significantly higher variability (t-test, $p > .001$) as more precipitation falls in the form of rain, entering the runoff system faster than does the snow falling in the glacierised basin.

The character of the hydrographs typically changes during the month of January, signifying the change from autumn rain-dominated hydrology to winter snowdominated hydrology. Winter typically demonstrates a decline in mean flow and runoff peak flows as precipitation accumulates in the snowpack. Both runoff peak flows and base flow decrease more in the glacierised basin than in the unglacierised basin.

For example, the peak flow from a storm on 24 January 1998 was 64.7 cubic metres per second in the unglacierised basin but only 38.4 cubic metres per second in the glacierised basin. The rain in the unglacierised basin increases the relaxation time of the recession limb of the runoff peak, allowing the precipitation to be converted to runoff and not allowing the flow to decline to a mean flow on the order of magnitude of earlier in the season. At the same time, the snow in the glacierised basin decreases the relaxation time of the recession limb of the runoff peak, resulting in less runoff and allowing the flow to decline to a mean flow on the order of magnitude of earlier in the season. Mean flow in the glacierised basin declines below summer flow levels, to the range of 11-17 cubic metres per second, while the unglacierised basin mean flow has dropped to 17-25 cubic metres per second. The coefficient of variation remains high in each basin because discharge is driven by random storm events.

The snowmelt season begins when the higher elevations start to warm up. This snowmelt may be seen in the hydrographs as a gradual rise in mean flow from March into June. The snowmelt is more pronounced in the higher- elevation, glacierised basin.

This season corresponds with the northward movement of the polar front and the subsequent decrease in the number and intensity of precipitation events and warming resulting from clearer skies. The onset of the snowmelt season begins in the glacierised basin when the mean flow rises in April or May from approximately 11 cubic metres per second and reaches a maximum during June at around 42 cubic metres per second, then gradually declines during July. The drop in mean flow during July represents the time when the snowpack is disappearing, leaving only patches of snow and ice in high-elevation locations.

In the unglacierised basin the snowmelt season is different. The snowmelt begins about two weeks earlier and reaches a peak about 70 percent the size of the glacierised basin about a month earlier (in mid-May). The decline in the snowmelt-season discharge is much faster, dropping streamflows well below the levels in the glacierised basin. The variability in daily discharge changes, from wide fluctuations around 1.0 to a stable level around 0.5, similar to the variability in the glacierised basin. The stabilisation in variability indicates a change from storm event dominated runoff to temperature-driven snowmelt runoff. A t-test ($p > .001$) indicates that differences in discharge between the glacierised and unglacierised basins during this season are significant. The mean flows in the glacierised are twice those in the unglacierised basin.

The summer low-flow season represents the warm, dry season of the Pacific Northwest, when the climate is dominated by the subtropical high, little precipitation fails, and the snowpack has largely disappeared. This hydrologic season is best illustrated in the unglacierised basin during the months of August and September, when flows drop to 6-8 cubic metres per second with few runoff peaks. During this season, in the unglacierised basin mean flow comes largely from groundwater discharge.

The mean flow in the glacierised basin declines to 14-20 cubic metres per second but exhibits more variability. The discharge of the glacierised basin is two to four times larger than that of the unglacierised basin as a result of cooler temperatures, more surviving snowpack, and inputs from glacier meltwater. The variability in the glacierised-basin discharges during this season results from the increased melting of snow and glacier ice during warm periods, rather than from precipitation. Few precipitati on-runoff peaks occur during a given year, for dry soil absorbs most of the rain.

The four seasonal runoff regimes can be seen clearly in the difference in discharge produced by the two basins. The autumn rainfall season is a transition between summer-flow and winter-flow regimes. The winter snowfall season is dominated by high day-to-day discharge variability. The unglacierised basin produces more runoff from precipitation in the form of rain because of its lower mean elevation; less runoff is produced in the glacierised basin, where snow is accumulating. The snowmelt season is clearly seen as the glacierised basin begins to contribute more discharge, reaching a peak in June and declining into August as the snowpack dwindles. The moderate discharge variability indicates the influence of localised rain events and warm days. The late summer low-flow season has little variability because few storms produce precipitation and because less snow and ice are available to melt.

Total Basin Flow Contributed By High-Elevation Snowsheds

The percentage of high elevation snowshed meltwater contributed to the lower Nooksack River from the glacierised basin is important to total basin runoff during the late spring, the summer, and the early autumn. Little or no high-elevation snowshed contribution is made to the main-stem Nooksack flow from the glacierised basin during the winter, and occasional contributions are made in the autumn and spring transitional seasons, when both rain and melting snow/ice are contributing to basin runoff.

The rise in runoff contribution from the glacierised basin in May occurs as the lower-elevation snowpack dissipates and the snow and glacier meltwater from the higher elevations of the glacierised basin begins to contribute significantly to runoff. From mid-July through August, about 30 percent of the mean daily discharge comes from high-elevation snowshed melt. The mean high-elevation snowshed meltwater contribution for the summer discharge equates to 6.2 percent of the annual discharge.

The interannual variability in the percentage of glacier and/or snow meltwater contributed by the glacierised subbasin represents differences in snowpack accumulation and energy available in the high-elevation region. The annual contributions are normally distributed, ranging from 15.7 percent (in 1972) to 39.9 percent (in 1950) of the seasonal discharge. Interannual and secular variability in high-elevation snowshed contribution is considerable but increased from an average of 25.2 percent in the 1940s to an average of 30.8 percent in the 1990s. High-elevation snowshed meltwater contributions reach a peak in average daily contribution of 32.7 percent on 12 August, with average daily contributions of more than 30 percent between 22 July and 20 August, a total of thirty days. Individual years show considerable variability, with contributions exceeding 30 percent occurring as early as 10 June (in 1941) and as late as 8 October (in 1999) and lasting up to 118 days in 1999 but not reaching the 30 percent level on any day in 1950 or 1956. Peak daily high-elevation snowshed meltwater contributions have been as high as 63.9 percent (on 19 July 1995) and as low as 27.9 percent (on 5 September 1955).

Evaluating Estimates

Estimating high-elevation snowshed contributions to streamflow is essential, because direct measurements from discharging glacial sources are lacking. My approach takes advantage of a unique geographical setting where discharge data from similar side-by-side glacierised and unglacierised basins can be compared. Although the approach has shortcomings discordance between the points of discharge measurement and the high-elevation snowshed sources; poor spatial resolution (35.6 percent of basin area and 37.2 percent of basin glaciers); no records of the diurnal variability of stream discharge my estimate improves on previous estimates by increasing temporal resolution from monthly averages to daily averages. Daily temporal resolution allows capture of significant variations that can be averaged away when using monthly discharge data. The average annual hydrographs show peaks in seasonal discharge that correspond with peak estimates based on energy.

The average estimated high-elevation snowshed meltwater contribution for the late summer (26.9 percent) agrees well with estimates of glacier meltwater contributions in the North Cascades. The estimates determined in my study represent a combination of glacier meltwater and high-elevation snowpacks, so they are maximum estimates for glacier meltwater. Studies of Mount Baker glaciers suggest that they were retreating prior to 1949, advanced between 1949 and 1978, and then began a gradual retreat. The same temporal trends can be observed in the high-elevation snowshed meltwater contribution data. The earliest years of record have average high-elevation snowshed contributions in excess of 30 percent of the streamflow, indicative of a high ablation rate. The contribution declines into the 1940S, signifying the end of the recession period, then remains consistently low through the early 1980s while the glaciers were accumulating volume and advancing. The retreat that began in the early 1980s is clearly expressed as the meltwater contribution increases from approximately 23 percent (average, 1975-1984) to approximately 31 percent (average, 1990-1999).

Mass-balance studies have been carried out on two glaciers just outside the watershed. The Lower Curtis Glacier is a cirque glacier on the western flank of Mount Shuksan, and the Rainbow Glacier is an outlet glacier from the Mount Baker ice cap. Correlations of mass-balance data between fourteen glaciers across the North Cascades indicate that the glaciers tend to fluctuate in concert with one another. The annual mass-balance measurements (1984-1998) have significant negative relationships with annual high-elevation snowshed meltwater contribution. The mass balance of the Lower Curtis Glacier has a stronger relationship than does that of the Rainbow Glacier. This expected negative relationship suggests that the estimate represents, to some degree, high-elevation snowshed meltwater contributions to the watershed.

Impacts of High-Elevation Snowshed Meltwater on Annual Hydrology

High-elevation snowshed meltwater has five principal impacts on watershed hydrology: lower stream temperatures; sudden, unexpected contributions to discharge caused by glacial reservoir outbursts; increased summer discharges; temporal delays in maximum spring discharges; and decreased annual, especially summer, variations in runoff. My study did not address stream temperature or glacial outbursts, but the summer increase in discharge, delay of spring peak discharge, and decrease in runoff variability were detected.

The mean high-elevation snowshed meltwater contribution in the Nooksack Basin for the summer discharge was estimated to be 26.9 percent. During the summer the glacierised basin produces about three times the amount of streamflow as does the unglacierised basin. This quantity of water may not seem overly important, but summers are dry and streams run low while human demand increases dramatically. In the absence of high-elevation snowsheds and glaciers, the summer discharge of the lower Nooksack River would drop, resembling the hydro graph of the unglacierised basin.

Maximum spring runoff occurs an average of twenty-nine days later in the glacierised basin than in the unglacierised basin, whereas its snowmelt season only starts seven days later. The delay in high-elevation snowshed melt is due to a slow rate of warming at higher elevations and to a larger volume of snow to melt. The quantity of meltwater in the unglacierised basin is limited by the snowpack volume, but the glacierised basin has a meltwater source that is limited only by the energy required for melting. The low-elevation snowpack generally disappears from the North Cascades by late June. The glaciers, however, continue to melt at an increasing rate as temperature, solar radiation, cloud cover, and snow-ice albedo factors combine for effective ablation through August.

The coefficient of variation of daily discharge is significantly (t-tests, $p > .001$) smaller during all seasons except winter and annually in the glacierised basin than in the unglacierised basin. The lower coefficient of variation during the autumn is due to more precipitation accumulating as snow in the glacierised basin than the unglacierised basin, as well as to extended snowmelt during warm years. The coefficient of variation converges between the two basins in winter as snowfall becomes the mode of precipitation. Both basins experience a significant drop in the coefficient of variation during the spring snowmelt as precipitation inputs to streamflow become less important. Although significantly higher in value, the spring unglacierised-basin coefficient of variation is reduced by the snowmelt, causing basin variability to behave in a glacier-like manner.

The greatest difference between the two basins is in summer. The glacierised-basin discharge is fairly consistent from day to day, fluctuating largely in response to temperature and solar radiation, which melt the snow and glacial ice. Precipitation events produce only modest increases (if any) in discharge because the accompanying cloud cover reduces melting and because the dry soil absorbs most of the rainfall. In the low-flowing, unglacierised basin the same precipitation event produces a much larger runoff. The dimensionless coefficients of variation allow comparisons to be made between watersheds of different sizes.

The coefficient of variation for the entire upper Nooksack Basin (Deming station) is not significantly different (t-test, $p > .0001$) from that for the glacierised subbasin on an annual or seasonal basis, except in winter. The coefficient of variation for the unglacierised subbasin is always significantly larger than the Deming coefficient of variation, however. Even though glaciers constitute only a small portion of the entire watershed, hydrologic processes associated with them have a distinct imprint on the downstream watershed, buffering daily variations in flow. The high-elevation snowshed meltwater contributions in summer and the accumulation of snow in high elevations of the basin reduce the variability of the surface runoff within the entire watershed.

My analysis of coefficients of variation appears to be at odds with the analysis of Andrew Fountain and Wendell Tangborn (1985), who found low a coefficient of variation in the winter and a high coefficient of variation in the summer. The difference stems from timing of the sampling: My study uses daily discharge data, whereas Fountain and Tangborn use monthly data. On a monthly basis, autumn and winter months do not exhibit much interannual variation (that is, rainfall is always abundant), whereas individual days show a tremendous variability (for example, no rain one year, but 5 centimetres the next). On a monthly basis, summer months exhibit considerable interannual variation, whereas individual days show less variability (that is, mostly with no rainfall). Ignoring the differences in magnitude, seasonal trends in the coefficient of variation are consistent.

Interannual variations in high-elevation snowshed meltwater are a product of seasonal temperature and precipitation regimes, glacier mass balance, and the health and size of glaciers. Alpine-glacier mass balance is quite sensitive to changes in synoptic climatology: Any meteorological processes operating over glacier surfaces must also occur over nonglacierised portions of highelevation snowsheds. In other words, when the snowpack is abundant or the temperatures are cold, less high-elevation snowshed meltwater is contributed to runoff, whereas during drought or warm years high-elevation snowshed meltwater augments streamflow. The net effect of the hydrologic impacts of high elevation snowsheds is to buffer the year to year variation in streamflow caused by the variations in precipitation. Glacierised basins will produce more meltwater in dry or warm years than in wet or cool years, thus reducing the interannual fluctuations in discharge.

Possible Effects of Climatic Changes

Glacier environments are especially sensitive to climate changes and variability. Several climate models agree that the Pacific Northwest will probably experience a 1.7-2.8[degrees]C temperature increase and a 1-10 centimetre increase in winter precipitation during the early twenty-first century. As a result, runoff characteristics may change appreciably over the next several decades. In the near term the effects of global warming are likely to be masked by ongoing year to year climatic variability. The relatively low elevation of much of the surface area of the North Cascades glaciers could cause their retreat and demise during this century. Higher temperatures will probably both cause a higher percentage of annual precipitation to fall as rain (that is, higher snow lines) and accelerate summer ablation. As glacier recession accelerates, summer runoff is likely to increase. The trend Seen in the Nooksack data over the last sixty years, where the contribution by the high-elevation snowshed has increased from approximately 25 percent to approximately 30 percent, will probably continue. After the glaciers have largely melted away, their input will be lost, resulting in a reduction in flow by 25-30 percent. This reduction in stream discharge is likely to occur within a few years near the end of a glacier's life.

Though not currently predicted, the climate of the North Cascades could become less conducive to high-elevation snowshed meltwater, a fear also expressed for other parts of the Cascades and even for the Sierra Nevada of California and Nevada. If annual or seasonal temperatures fall, snow and ice melting and, therefore, summer streamflows will decline. If snowfall increases, the high-elevation snowshed meltwater contribution will decrease because the glaciers will remain under snow cover for longer periods in summer. An increased snowpack will lead to an increase in summer streamflows, however, because more snow will be available to melt. As discussed above, the high-elevation snowshed meltwater contribution to the flow of the Nooksack River has fluctuated through time but appears to have increased in response to warming during the late twentieth century.

Regardless of climate changes, streamflow regimes and glacier mass balances in the Pacific Northwest are prone to interannual and secular variations related to large-scale patterns of climate variations over the Pacific, the El Nino Southern Oscillation and the Pacific Decadal Oscillation. The predominant pattern is that warm years tend to be relatively dry, with light snowpacks, lower spring streamflows, and higher summer streamflows, whereas cool years tend to be relatively wet, with heavy snowpacks and high streamflows. Warmer years tend to see water shortages, increased glacier melt, and less-abundant salmon runs. The Pacific Decadal Oscillation has been in the warm, dry phase since 1977 but appears to be moving toward the cool, wet phase. Global warming is likely to increase the frequency and magnitude of these climatic variations and their impacts on the hydrological system.

Implications for Water Resources

The high-elevation snowsheds in the Nooksack River watershed have a significant impact on downstream hydrology. Although the influences of high-elevation snowshed meltwater fluctuate from year to year with meteorological conditions and trends, the hydrology of the Nooksack Basin strongly reflects glacial conditions. Water users in the basin, including its ecosystems, rely on the seasonal patterns and expected range of variation in surface runoff produced by the glacierised watershed. Two modes of change are at work within this watershed, however. Rapid population growth and rising environmental concerns are increasing demand for water. Climate change and variability appear to be augmenting the surface supply by melting glaciers, so the higher rates of flow evinced in glacier-fed basins are artificially and unsustainably high; when the glaciers go, so will the water supply. If current climate trends continue, when the glacier mass has entirely melted, the shortfall in the overappropriated basin hydrology will be significant. A new hydrologic regime will be in place: 25-30 percent less summer flow, spring runoff peaks earlier in the year, and increased variability in streamflow, including larger storm-peak flows.

Rapid population growth, increasing environmental concerns, and resulting changes in the character of water demands have led to increased competition for water even under normal flow conditions. Western water-management practices, storage infrastructure, and patterns of use are tuned to the expected range of variation in surface runoff and groundwater availability. The abundant surface water supply in the upper Nooksack Basin has promoted a historic reliance on this resource throughout the entire basin. Growing human and environmental demands on water quantity and quality require comprehensive water-development strategies that incorporate the relationship between lowland and upper watershed supplies. My study suggests that late-summer surface supplies are derived, in part, from a limited source: glaciers. Because glacier volume and high-elevation snowfall respond to climate change and variability, the input from these sources is likely to change during this century.

The Nooksack River supplies drinking water, hydroelectric power, recreation, fisheries habitat, irrigation water, and water for several industries in the region (Benjamin and others 1999). Demands increase substantially during the summer. Were water supplies to reduced by

20-30 percent, the system would further tax the slim margins on which it operates today. Water withdrawals for human use would need to be reduced, resulting in economic hardships in the agricultural, industrial, and eventually service sectors (Rodda 1995). Indeed, this scenario was witnessed in California in the late 1980s and early 1990S following multiple years of drought (Harding, Sangoyomi, and Payton 1995).

Fish populations and other aquatic resources are affected by changes in seasonal flow regimes, total flows, warmer water temperatures, and associated changes in water quality. Changes affect the health of aquatic ecosystems, with impacts on spawning practices, productivity, species diversity, and species distribution. Salmon populations rely on the Nooksack River for various distinct habitat requirements during different stages of their life cycles. Hydrologic changes to their riverine habitat may further negatively affect their already severely depressed populations.

Clearly, high-elevation snowsheds are important to the regional water supply. Recognising and predicting further changes in water supply will help water managers provide water to the growing population of the region while protecting already heavily impacted aquatic ecosystems. New reservoirs and water-transfer systems require considerable lead time to plan and construct. Such structures maybe needed to deal with declining water supplies in this part of Washington. It is critical that we understand how water is supplied today so we can manage the resources under different climatic conditions that may exist in the future.

These changes will have an impact all water users in the basin. New allocation and distribution systems are required to meet the new water regime. Additional flood controls and storage facilities may be required if variability in flow increases. Hydroclimatic variability represents a major challenge for land and water-resource development in the watershed. My study increased our understanding of how high-elevation regions influence a lowland water supply. Effective water-development planning and policymaking must recognise how changes in upper-watershed conditions will affect water resources.

Chapter 5

New Generation of Indicators for Coastal Waters

More than half the world's human population lives within 100 km of the coast, and that number is expected to increase by 25% over the next two decades. Consequently, coastal ecosystems are at serious risk. Larger coastal populations and increasing development have led to increased loading of toxic substances, nutrients and pathogens with subsequent algal blooms, hypoxia, beach closures, and damage to coastal fisheries. Recent climate change has led to the rise in sea level with loss of coastal wetlands and saltwater intrusion into coastal aquifers.

Coastal resources have traditionally been monitored on a stressor-by-stressor basis such as for nutrient loading or dissolved oxygen. To fully measure the complexities of coastal systems, we must develop a new set of ecologic indicators that span the realm of biological organisation from genetic markers to entire ecosystems and are broadly applicable across geographic regions while integrating stressor types. We briefly review recent developments in ecologic indicators and emphasize the need for improvements in understanding of stress-response relationships, contributions of multiple stressors, assessments over different spatial and temporal scales, and reference conditions.

We provide two examples of ecologic indicators that can improve our understanding of these inherent problems: a) the use of photopigments as indicators of the interactive effects of nutrients and hydrology, and b) biological community approaches that use multiple taxa to detect effects on ecosystem structure and function. These indicators are essential to measure the condition of coastal resources, to diagnose stressors, to communicate change to the public, and ultimately to protect human health and the quality of the coastal environment.

More than half the world's human population resides within 100 km of the coastline [National Oceanic and Atmospheric Administration], with increases likely over the next two decades. The coastal zone represents at least half the value of global ecologic services, and in economic terms, is the single most important source of recreational and residential income worldwide as well as fisheries.

Human development of coastal watersheds has greatly accelerated environmental pressure on downstream estuarine and coastal ecosystems; yet, unfortunately, assessing detrimental changes in these systems is complex. The symptoms of degradation include deterioration of water quality, loss of habitat and biodiversity, beach closings, fishery declines, fish consumption advisories, and an overall decline in the livability in the coastal zone. Coastal systems are hydrologically complex and are among the most susceptible to direct disturbance through global climate change. For instance, sea levels have been rising over the past century and even greater rises are predicted over the next 50 years. These changes will continue to affect coastlines and will dramatically increase saltwater intrusion into freshwater coastal aquifers as well as the displacement of coastal agriculture. Climate change is also believed to have increased events of heavy precipitation and flooding, which recently have become more common. These events increase the flushing of nutrients and toxic chemicals into coastal regions. In addition, many estuaries are highly urbanised and reference conditions are difficult or impossible to specify because of large changes in habitat, water diversion, and the introduction of exotic species.

Not only are estuarine systems complex, but many stressors are also difficult to control. For example, non-point source nutrient pollution from coastal watersheds is a major problem, that has affected more than 60% of coastal rivers and bays. Increases in nitrogen and phosphorus coming into coastal ecosystems have led to disruptions of basic ecologic functions [e.g., rising frequency and proliferation of harmful algal blooms (HABs) and increasing oxygen depletion (hypoxia)] with major damage to coastal fisheries and biodiversity. In addition, nonpoint sources of toxic substances (e.g., agricultural chemicals and urban runoff) have impaired habitat quality for aquatic life and the human use of numerous coastal watersheds.

It is increasingly evident that ecosystem and human health are intricately linked. For example, HABs can cause diseases in humans that result from consumption of contaminated seafood or inhalation of toxins entrapped in sea spray. Moreover, the distribution and frequency of HAB events have increased along U.S. coastlines over the last 30 years. People are also exposed to waterborne diseases through recreational contact, and the incidence of these diseases is increasing worldwide, as is the cost to the economy of frequent beach closures. The principal agents of these diseases are bacteria, viruses, and protists. In the case of bacteria, this includes both native marine organisms (e.g., Vibrio species) and human or animal-derived pathogens from sewage and runoff. The importance of these connections between coastal conditions and human health was realised by the National Institute of Environmental Health Sciences and National Science Foundation (Washington, DC), and they subsequently established their Centres for Oceans and Human Health.

Given the importance of coastal systems and the increasing pressure on them, quantitative measures of coastal ecologic conditions are absolutely essential for detection of change as well as for the design of control measures and restoration activities. Many approaches to comprehensive assessment of condition are cost prohibitive; thus, there has been a tendency to use broadly applicable indicators such as water clarity, nutrient or contaminant loads and levels, and various biodiversity measures such as species richness as metrics. Yet measurements of environmental condition are becoming more sophisticated and more applicable across space, time, and biological organisation. These new indicators include such metrics as diagnostic photopigments of algal functional groups to assess eutrophication; biochemical and genetic indicators of toxicant exposure and stress; molecular techniques to assess human fecal bacterial distribution; isotopic techniques to evaluate nutrient enrichment; indices of biological integrity or other biological community responses; ecosystem and population modelling approaches; landscape metrics; and remote sensing techniques to detect large-scale land use impacts and change. While these indicators represent impressive advancements in both science and technology, there are limitations on their widespread and integrated use (NRC 2000).

Our primary goal in this review is to identify four critical areas in which scientific advancements are needed before improvements can be made in indicator development of coastal regions. In addition, we provide examples of two promising approaches to improvements in indicators. Defining the limitations of previous approaches and developing new approaches was a major goal of the U.S. Environmental Protection Agency (U.S. EPA) Science to Achieve Results Program (STAR) in establishing the Estuarine and Great Lakes (EaGLe) Research Program. The EaGLe investigators (Atlantic, Pacific, Gulf of Mexico, and the Great Lakes coastal areas) realise there is increasing urgency to develop indicators capable of detecting and diagnosing environmental conditions over space and time at cellular, organism, habitat, ecosystem, and regional levels. As the explosion of technical and conceptual advances in various disciplines ranging from molecular biology to ecosystems ecology and from remote sensing to bioinformatics continues to provide new and better tools, the goals of the EaGLe program and the science surrounding them are works in progress.

Limitations of Current Coastal Indicators

Most indicators were designed to provide specific information on local conditions such as water clarity or eutrophication or to provide broad-based snapshots of regional-scale water quality and habitat condition as exemplified by the total maximum daily load or Environmental Monitoring and Assessment Program (EMAP). From these studies we know that large areas of the U.S. coastal zones are impaired, and many of the specific stressors within coastal watersheds that contribute to impairments have been identified. Traditionally, the focus has been on indicators associated with water quality and toxic substances; however, there are major concerns on many other stressors, including habitat and landscape change, exotic species, global climate change, and pathogens. We focus on some of the limitations of current indicators that must be overcome for future advancements to be made on broad aspects of this problem in the coastal region.

Stress with response. Most current indicators of coastal condition are not linked with specific stressors; hence, it is unclear what causes the change in the indicator and, therefore, what management solutions should be implemented to affect an improvement.

Multiple stressors. Stressors in coastal ecosystems are diverse and originate from both anthropogenic and natural perturbations. Most current indicators are incapable of providing diagnostic information and are unable to discern the relative contribution of the various stressors to the observed changes in the indicators. Space and time. Sources of stress to coastal environments operate over a range of spatial scales (e.g., square metres to entire landscapes) and time (seconds to decades). Current indicators are typically not explicit in how they relate condition with stressors over these different scales.

Reference conditions. The interpretation of the condition or change of an indicator is based on a comparison to reference conditions or benchmarks. Frequently, these reference conditions are not quantitatively defined; thus condition or meaning of a change in indicators is subject to considerable interpretation and debate.

Linking Stress with Response

From a management perspective, among the greatest limitations of many indicators of coastal condition is the lack of a link with the cause for change. In the development of indicators, distinguishing between measurements of disturbance or stress and measurements of ecosystem response is imperative. The terminology in this area of the literature varies considerably, but recent reviews provide clarification.

In developing stress response relationships, natural experiments or surveys across a disturbance gradient from relatively pristine to highly disturbed areas are used most often. In these situations, it is often difficult to control the stress because multiple factors often vary across any environmental gradient. Thus, inherent limitations in the scope of inference result and must be explicitly recognised. Understanding the response variable is also dependent on the strength of conceptual models used to describe the factors structuring the ecosystem, and on the extent to which anthropogenic disturbances influence that ecosystem. Obviously, detection of a response to a stressor is best accomplished by experiments in which the stressor can be manipulated in frequency, intensity, and duration. Unfortunately, these experiments often lack realism, as they are typically limited to laboratory situations or small-scale field-based mesocosms. However, a combination of a gradient design with field and laboratory experiments can be a powerful approach for the initial phases of indicator development, but as the indicator is applied at larger spatial scales or in uncharacterised sites, additional approaches are needed. Multivariate statistical approaches, novel modelling approaches, and techniques to aggregate indicators according to their use in management may all be valuable approaches. For example, these may permit more realistic diagnosis of stressors as complex as nutrient inputs, pathogen effects, and toxicant bioaccumulation. A number of existing approaches are available to couple toxicant stress with response. For example, evaluation of the effects of toxic substances on ecosystems can involve multiple approaches: a) comparison of a toxic response or specific dose level of a contaminant to a water quality standard that has been linked to biological effects; b) assessment of environmental effects of multiple toxic substances by using well-characterised individual contaminant exposures; or c) the coupling of physiologic and genetic indicators with environmental chemistry and ecologic responses at multiple spatial and temporal scales.

However, these too have their limitations. The first approach is limited when multiple contaminants and multiple stressor types are present. The second approach is primarily limited because often no direct integration of the toxic response and exposure occurs. The result is data correlation, which is also limited to the spatial and temporal scopes of the immediate investigation. The third approach considers multiple stressors and permits direct integration and scaling, but significant challenges remain to fully develop indicators for a range of habitats and model organisms. There are also combinations of these approaches. A simple index of sediment contamination, the Sediment Quality Triad, combines the first two. It provides a framework for analysing benthic community data, analytical chemistry, and toxicity test data to assess whether a site is affected by toxicants, and is widely used throughout the nation. Yet, the action levels derived for specific contaminants are often unknown and benthic data are highly variable. Moreover, there is often a lack of specific reference conditions that precludes dear interpretation. The triad approach is also based on acute lethality, whereas sublethal effects can also be very important.

In summary, there is a need for controlled experiments in laboratory and field settings for those stressors (e.g., toxicants, nutrients, turbidity) amenable to manipulation. For larger-scale stressors such as introduction of exotic species, climate change, habitat loss, or landscape change such as fragmentation that are not easily amenable to manipulation, field experimental designs that test responses over gradients of stressor levels are among the options for linking stress with response (Danz et al., in press). Coupling the approaches of laboratory and field methods are essential for future development of appropriate response indicators.

Multiple Stressors of Environmental Condition

Stress on coastal ecosystems is usually a combined effect of natural and anthropogenic disturbances. Natural disturbances in the U.S. coastal zones include water-level fluctuations from droughts and floods, wind events such as hurricanes, natural soil/sediment deposition, insect infestations, and forest fires. These natural disturbances vary in intensity both spatially and temporally. Major anthropogenic disturbances to the watershed of coastal ecosystems include permanent land cover conversions of native vegetation to agricultural, residential, and industrial areas; and temporary conversions of land due to forestry.

These conversions result in concomitant disturbance to coastal ecosystems, including a) landscape effects of fragmentation, b) increased surface water runoff, c) increased nutrient and sediment input, d) increased pesticide and other chemical inputs, e) increased water temperature, and f) greater human disturbance from recreational use, increased fish and shellfish extraction, and noise. Climate change and the resulting change in weather patterns is a combination of both natural, stochastic events, and human-induced warming which affects vegetation, water levels, and virtually all types of disturbance. In arid regions, water diversion and water use patterns also result in landscape and ecosystem-level alterations. Detecting the effects of both individual disturbances and the simultaneous influences of natural and anthropogenic disturbances in coastal ecosystems is a challenging and complex task.

The relative effects of anthropogenic disturbance must be distinguished from the ranges of variation in natural disturbance regimes, but because of the large size and variability of coastal ecosystems, manipulative experiments to untangle the complexities of the varying disturbance regimes are difficult except on a relatively small scale. Combining specific indicators with modelling efforts can clarify and distinguish anthropogenic from natural stress in individual ecosystems and regions. However, the general trend has moved away from using direct diagnostic measures of stressors to using integrated indicators of ecosystem structure and function. Yet, we know this approach may be inadequate for many applications because stressors vary among regions, implying that indicators are needed with diagnostic and prognostic capabilities. Indicators, therefore, could be grouped to fit regional needs related not only to assessing condition but also to developing appropriate management responses.

Characterising the effects of multiple stressors on any ecosystem is among the most challenging tasks facing scientists today because multiple stressors can have synergistic, additive, or antagonistic effects on biological responses. Disentangling the various effects of multiple stressors will likely require a combination of controlled laboratory experiments, large-scale studies over multidimensional gradients of stress, and insightful modelling of ecosystem responses and change.

Spatial and Temporal Explicitness

Ecologic indicators are constructed or selected to assess the condition of ecosystems and to detect environmental change related to human disturbance. Condition is often assessed by documenting the state or rate of ecologic processes such as productivity, respiration, or the structuring of biological communities. Indicators may do this by either measuring those processes directly (such as primary productivity) or inferring process from pattern (such as indices of biotic integrity (IBIs) as descriptors of community structure). Ecologic processes operate over a range of spatial and temporal scales, and the resulting patterns are expressed over varied scales. Hence, the relevant scale of each indicator must be specified to relate pattern to process in the appropriate conceptual model. Levin (1992) stated "the concepts of scale and pattern are ineluctably intertwined. The description of pattern is the description of variation, and the quantification of variation requires the determination of scales."

Many studies have sought to quantify spatio temporal patterns across a range of scales. Unfortunately, few have determined whether patterns are consistent across scales or whether related phenomena cross scales. In addition, because of technical, logistical, and financial reasons, most ecologic studies have focused on small systems (e.g., site or plot level) and short periods of time, which in turn has limited the development over large spatial scales. Alternatives such as top-down approaches do exist. Here, large-scale processes form the basis of inferring process from pattern and are being applied in regional classification schemes.

The need for scale explicitness is complicated by multiple stressors arising from human and natural disturbances. Most ecologic indicators are related to multiple stressors and scales. For example, in a study of littoral macroinvertebrate communities, 23% of the variance in taxonomic composition was associated with habitat factors, but greater spatial scales (riparian, catchment, ecoregion classification) accounted for 24% of the variance. If indicators are to be used effectively in management, it is necessary that we know the relevant scale(s), so that the scale of management actions matches the scale of the phenomena being measured. Experimental approaches that allow for the partitioning of the variance among different stress components and over a hierarchy of spatial scales will be critical in the development of new ecologic indicators.

Reference Conditions

To interpret any set of indicators, one must compare results of monitoring to standards or benchmarks. One of the preferred benchmarks is a reference site or condition. The use of reference sites has become increasingly common as ecologists and managers search for reasonable and scientifically based methods to measure and describe the inherent variability in natural aquatic systems. As there are likely few places on earth unaffected by anthropogenic disturbances, true reference areas remain elusive. For example, even coastal regions of Greenland or Antarctica have been affected by atmospheric chemical inputs and climate change. Alternatively, even in coastal regions with long histories of human occupation and, hence, anthropogenic disturbances, reference sites can be established using specific estuaries, watersheds, or lotic systems entering the coastal zone. These may represent the best attainable environmental conditions for a specific geographic setting, a historic representation using paleolimnologic data, a simulated reference condition, or a situation where conditions fall within the range of natural variability for the system.

Determining reference condition of a system is highly dependent on the indicators used and the locations where samples were gathered. Benthic indicators will provide different results than fish indicators. Similarly, indicators will be different in large, ephemerally stratified systems compared with smaller, well-flushed systems. For example, phytoplankton growth responses to nutrient enrichment will not be as profound as those for benthic microalgae in well-flushed systems. Here, benthic microalgae may be more sensitive and meaningful indicators of ecosystem response to nutrient enrichment. Indicators of community structure (i.e., diversity indices, keystone species) may gauge ecosystem conditions quite distinct from indicators of function (e.g., primary and secondary production, respiration, and nutrient cycling). IBI, habitat suitability indices, and chemical monitoring are specific examples of indicators that in combination can assess structure, physical chemical quality, and biological measures of reference condition.

Historical information from a site, such as survey data, paleolimnologic studies, and habitat reconstruction, can be extremely helpful for determining reference conditions. Finding such information, however, can also be very difficult and most historical information is not quantitative such as for urban estuaries. Hughes (1995) summarised the basic characteristics necessary for a suitable reference condition including reasonableness and political acceptability, sufficient number of reference sites within the area of interest, and suitable data on natural conditions of the site.

Examples of New Indicators

Environmental indicators can have an enormous number of possible end points, reflecting the breadth and diversity of the scientific underpinnings in biology, chemistry, and physics. For example, biological indicators span the realm of biological organisation from genetic markers to entire ecosystems. Chemical indicators reflect a variety of spatial or temporal scales ranging from oxygen demand for a specific point source to global carbon dioxide distributions in the atmosphere. Physical indicators can include elevational, morphologic, transport, circulation, exchange and stratification processes with all their attendant ramifications for ecosystem structure and function. Because of the massive amounts of information that can be gathered across levels of physical, chemical, and biological organisation and across spatial or temporal scales, the challenge to integrate data across levels of organisation in space and time is daunting.

Current programmatic and academic funding scenarios exacerbate our lack of integration. Most funding is limited by amount and duration. These monetary and time deficiencies have been recognised by academic funding sources, such as NSF and U.S. EPA (EaGLe). As such programs mature, advances in integrating across a variety of trophic levels as well as organisational or spatial and temporal scales will likely occur.

We provide two brief examples of new types of indicators; one that links productivity (function) and hydrology and another linking community (structural) patterns. We believe these new types of indicators will substantially improve our ability to measure and understand the complexity, response, and condition of coastal systems. For example, analyses of photopigments provide a means to explicitly link nutrient and hydrologic stressors with specific phytoplankton groups and over explicit spatial scales when combined with remote sensing information. If data are gathered systematically over time, then temporal changes can also be linked with specific stress events (e.g., hurricanes). In the second example, multitaxa types of approaches provide a wide range of possibilities for improving our knowledge of stress response relationships, the identification of multiple stressor effects, spatial and temporal explicitness, and the identification of suitable reference conditions. For example, Luoma et al. (2001) provide eight case studies that specify cause and effect linkages using community analyses of macroinvertebrates to decipher the effects of multiple stressors. Paleolimnologic data derived from sediment and water column sampling of diatom communities are among the most powerful techniques for identifying suitable reference conditions within aquatic systems.

Photopigments as Integrators of Estuarine Nutrients and Hydrology

Nitrogen availability most frequently controls microalgal and higher plant primary production in estuarine and coastal waters. Loading rates of this nutrient directly reflects human population density and activity in coastal water and airsheds. Excessive nitrogen loading is a key causative agent for accelerating primary production or eutrophication. Symptoms include phytoplankton blooms, which may accumulate as ungrazed organic matter in the sediments, providing the "fuel" for oxygen consumption and depletion in bottom waters and sediments. This chain of events is particularly problematic in salinity- or temperature-stratified waters, where oxygen may not be easily replenished from the atmosphere. Hypoxic conditions alter nutrient cycling and promote fish disease and mortality.

Suspended microalgae or phytoplankton account for the bulk of estuarine and coastal primary production in many estuarine and coastal ecosystems. Their composition and activity are key in determining fertility, eutrophication, and water quality. Water discharge controls transport of phytoplankton through these systems and plays an interactive role with nutrient supply to control phytoplankton growth, competition, succession, and community composition. For example, high rates of freshwater discharge reduce the salinity and residence time. These conditions favour fast-growing oligohaline phytoplankton, such as chlorophytes (green algae). In contrast, low-discharge conditions promote long water residence, high salinity conditions, which favour slower growing halophylic taxa such as dinoflagellates and certain cyanobacteria. Phytoplankton community composition affects the structure and function of estuarine food webs, nutrient cycling, habitat condition, fishery resources, and overall ecosystem condition.

Chlorophyll a has been used for many years as a sensitive indicator of phytoplankton biomass. However, because virtually all phytoplankton contain this pigment, it alone cannot be used to determine community composition. Using additional diagnostic chlorophyll and carotenoid photopigments as indicators of major phytoplankton functional groups (i.e., diatoms, dinoflagellates, chlorophytes, cyanobacteria, cryptomonads), we can examine the interactive effects of nutrient and hydrologically driven changes of phytoplankton community composition and activity. HPLC, coupled to photodiode array spectrophotometry is used to determine phytoplankton group composition based on the diagnostic photopigments. Photopigment markers include chlorophyll b and lutein (chlorophytes), zeaxanthin, myxoxanthophyll, and echinenone (cyanobacteria), fucoxanthin (diatoms), peridinin (dinoflagellates), and alloxanthin (cryptomonads). A statistical procedure, ChemTax, partitions chlorophyll a (i.e., total microalgal biomass) into the major algal groups to determine the relative and absolute contributions of each group.

Examples from ongoing studies in the Neuse River estuary in North Carolina and Pamlico Sound (1994-present) show that these systems have experienced the combined stresses of anthropogenic nutrient enrichment, droughts (reduced flushing combined with minimal nutrient inputs), and since 1996, elevated hurricane activity (high flushing accompanied by elevated nutrient inputs). Seasonal and hurricane-induced variations in river discharge, and the resulting changes in flushing rates, and hence, estuarine residence times, have differentially affected phytoplankton taxonomic groups as a function of their contrasting growth characteristics. For example, the relative contribution of chlorophytes, cryptophytes, and diatoms to the total chlorophyll a pool was strongly controlled by periods of elevated river flow. These effects are due to the efficient growth rates and enhanced nutrient uptake rates of these groups. Cyanobacteria, on the other hand, showed greater relative biomass when flushing was minimal (i.e., longer residence times) during the summer.

Further evidence that hydrologic changes have altered phytoplankton community structure is provided by the observed historical trends in dinoflagellate and chlorophyte abundance. Both decreases in the occurrence of winter-spring dinoflagellate blooms and increases in the abundance of chlorophytes coincided with the increased frequency and magnitude of tropical storms and hurricanes since 1996. The relatively slow growth rates of dinoflagellates may have led to their reduced abundance during these high river discharge events, These changes in the phytoplankton community have been linked to altered trophodynamics and nutrient cycling, which subsequently affects fishery habitats and yields.

Diagnostic photopigment analyses are able to detect significant changes in phytoplankton community composition over a broad range of time scales (< 24 hr to decades) and as such are well suited for monitoring programs designed to assess short- and long-term trends in water quality in response to: hydrographic features (circulation, upwelling); nutrient enrichment; climatic; and hydrologic perturbations (floods, droughts). In addition "top down" effects of grazing have been examined using an HPLC-based technique. Finally, these analyses have proven useful as a means of ground-truthing and calibrating remotely sensed estimates of phytoplankton bloom events. This coupling of indicator technology with remote sensing enables "scaling up," namely, mapping the spatial distributions of phytoplankton groups over large geographic areas not amenable to routine field sampling, evaluating the effectiveness of nutrient management strategies, use as an early warning system for blooms of nuisance or toxic species, and as a sensitive bioindicator of overall water quality conditions.

Biological Community Responses as Condition and Change Indicators

Plant and animal community structure and function have been extensively measured to describe the condition of both aquatic and terrestrial systems. The major challenge is how to scale and aggregate the responses of species populations at the site level to reflect conditions of the biological community level for specific taxa or to provide assessments of large scale patterns, such as IBI, biological species profiles, multitaxa indices, or indices of environmental integrity. These approaches hold tremendous potential for assessments of environmental conditions over large landscape or regional areas, as well as for detection of temporal change. However, these indicators will also require considerable development in the areas of a) providing linkages with single and multiple stressors; b) exploration of analytical techniques to integrate and synthesize multiple biological signals from species or functional groups within the biological community; c) parsing these multivariate responses among stressors and over varying spatial scales; and d) providing explicit spatial or temporal scales for the indicators which are consistent with the scales of management actions.

The strength of the community approach lies in the differential sensitivity of individual species, functional groups (e.g., guilds), or trophic levels to different stressors. Each of these levels can respond differently to stress. For example, O'Connor et al. (2000) found a correlation among many taxa (diatoms, benthos, zooplankton, fish, birds) to the gross condition of lakes, but fish provided the best measure of condition in the near-shore environment. In an experiment on the pesticide effects of mosquito control agents in wetlands, of the zooplankton, aquatic insect, and bird communities studied, only the aquatic insects exhibited a response to treatment. In this case, aquatic insects were the best indicator of pesticide effects. Niemi and McDonald (in press) provide many examples of responses by different taxa to diverse stressors.

The unique aspect of the community approach is the ability to sample a wide variety of taxa; each of which has a unique life history capable of being disrupted by stress at various scales. All coastal regions are represented by thousands of species including taxa such as bacteria, plankton, macroinvertebrates, fish, vascular and nonvascular plants, amphibians, and birds. Many associations between these taxa and stress exist. For example, diatoms are particularly sensitive to nutrients, whereas benthic invertebrates are responsive to sediment contamination in both lakes and estuaries. Fish communities are sensitive to human development and exotic species. Wetland vegetation is directly affected by hydrologic modifications such as dikes and road building. Amphibians are sensitive to water quality in wetlands. Bird populations are affected by landscape-level habitat change and fragmentation. Moreover, many of these taxa have well-established sampling methods, and some have long-term nationwide monitoring programs that are currently still in use. The combination of species- or taxa-specific responses by plant and animal communities to stressors and the availability of extant monitoring programs allows for the partitioning of multiple stress-response relationships.

Probability-based and standardised sampling of communities within specific sites but over large landscapes has proven useful for regional-scale assessments of environmental conditions. For example, these approaches have been used for the identification of imperiled systems; development of biological indicators for the Mid-Atlantic Highlands; and establishing the condition of streams and estuaries through U.S. EPA's EMAP. One important aspect of these large-scale approaches is the development of indicators that can identify areas that have the most severe problems and greatest need of management attention, action, and potentially restoration. Integration of these types of data will be challenging and will require multivariate, integrative approaches of multitaxa biological communities over large-scale landscapes and regions.

Promising new techniques to achieve integration of community measurements over multiple spatial, temporal, and biological scales are evolving such as development of multimetric indices, and statistical techniques. These techniques will require coupling with population and ecosystem-based models for aid in the interpretation of stressor risk and alternative management actions. Several large-scale programs have been initiated such as the EaGLe program reported here. For example, Danz et al. (in press) developed a stratified experimental design for the development of environmental indicators in the Great Lakes coastal region. This design was based on the compilation of more than 200 data layers on stress information for 762 coastal units and the identification of gradients of stress for several coastal ecosystem types. Six different taxa (amphibians, birds, diatoms, fish, macroinvertebrates, vegetation) were randomly sampled across these stress gradients to detect differential responses. This work is still in progress but identifies the tremendous potential of spatially explicit public databases in the future development of environmental indicators. With the exponential increase in technological and computational capabilities including molecular techniques, remote sensing technology, modelling and statistical sophistication, data management and storage, and internet communication, analysis of biological communities as indicator signals of environmental condition and change will rapidly advance.

Coastal ecosystems have experienced tremendous stress from natural and anthropogenic influences for hundreds of years, and stress levels are projected to increase in the future. A new generation of ecologic indicators is needed to measure the condition, diagnose stressors, communicate condition to the public, assess potential future status, and evaluate management actions in our coastal regions. Among the many challenges, the development of these indicators will require improvements in our scientific understanding of stress-response relationships, relative contributions of multiple stressors, how stressors operate over different organisational and spatio-temporal scales, and how reference conditions are determined. Fortunately, there is an explosion of ideas and technology that can aid in the multidisciplinary advancement of indicators in our coastal waters.

Chapter 6

Water Rights: The Water Policy Implications

Like many rivers throughout the western United States, the Columbia has been dramatically altered by human activities. A series of major dams and diversions have radically changed the big river and its tributaries, such as the Snake, the Yakima, and the Deschutes, turning them into an economic engine for the Pacific Northwest. So thoroughly has it been exploited for hydropower, navigation, and irrigation that the Columbia has been described as a river that has "died and been reborn as money." As the Columbia River system grew more industrialised, however, its legendary salmon populations declined sharply. The Columbia's salmon runs may once have numbered 16 million fish, but by the 1990s they had fallen to something like one million, and most of those fish were artificially produced in hatcheries. As more Pacific salmon populations were added to the national list of threatened and endangered species, recognition grew in the Pacific Northwest that the Columbia Basin ecosystem may have been pushed too far, and that changes would be needed to restore the salmon runs.

The Columbia, however, still faces new demands for water from farmers, cities, and others who continue to regard the River as a viable source. These new demands are sizable pending applications for new permits in Washington alone total up to 1.3 million acre-feet of water. These new demands could be viewed with some validity as either an incremental increase in use that is small in the context of the Columbia's annual flow, or as a new depletion that would further reduce river levels in the summertime when salmon are already stressed by low flows and high water temperatures. Irrigators and other would-be water users have argued strongly for the former view, while environmental groups and other salmon advocates have forcefully advocated the latter position.

Caught in the middle is the Washington Department of Ecology (Ecology), the agency responsible for decisions regarding new permits to withdraw water from the Columbia for use in Washington. Seeking a definitive scientific answer to the dispute over the potential impacts of new water withdrawals from the Columbia, Ecology asked the National Academy of Sciences (NAS) to study the issue. NAS, a nonprofit group of research scholars that is often called "The Supreme Court of Science," has issued a variety of influential reports on water management and the needs of native fish species, including a much-publicised 2002 draft report that questioned the scientific basis of Klamath Basin water management. Ecology's request was unusual, however, in that it came from a state agency, whereas NAS normally advises the federal government.

NAS released its report, Managing the Columbia River." Instream Flows, Water Withdrawals, and Salmon Survival (Report), on March 31, 2004. The Report finds that additional water withdrawals from the Columbia would, indeed, increase risks to salmon during critical periods of relatively low flows and high water temperatures. These risks are exacerbated because the future of the Columbia River system contains so many big uncertainties uncertainties regarding salmon survival, water supplies (both year to year and long-term), and water demands throughout the Columbia Basin. With its finding that new water withdrawals would increase salmon risks, the Report essentially says there is no such thing as a free drink, even from the mighty Columbia.

The Report does not stop there, however. In response to Ecology's request for comment on various scenarios for Columbia River water management, NAS made a number of findings and recommendations regarding potential new water withdrawals from the Columbia. Briefly stated, the Report urges a new approach to water permitting decisions: review of new permit applications by a multijurisdiction Columbia Basin water forum, water markets and conservation as the first option to supply water for new uses, tougher requirements to obtain new permits, and greater flexibility in water allocation and management. These recommendations may seem modest in an era when salmon recovery is a major priority of the Pacific Northwest, but they are contrary to long-standing traditions of water allocation and management in the western states. Moreover, the NAS recommendations carry major water policy implications for the

Columbia Basin and beyond, because the issues and challenges facing Ecology on the Columbia are not so different from those facing water managers throughout the West.

This Chapter identifies and briefly discusses the NAS Report's water policy implications. The point is not to analyse the Report and determine if its findings and recommendations are "correct," nor to provide a detailed evaluation of the Columbia River permitting controversy and the policy response to the Report. Instead, this Chapter focuses on the NAS recommendations regarding water allocation and management, explains how they differ from the traditional approach of western states, and discusses how water management might change if the states were to adopt these recommendations. In other words, the Chapter assumes the NAS recommendations constitute "good science," and assesses how the policy toward new water demands might change if states in the West were to see the NAS recommendations as relevant and implement them.

Water Development and Management on the Columbia

The Washington controversy is fundamentally a dispute over the best use of the waters of the Columbia River. These waters have sustained salmon for millions of years and the human inhabitants of the Columbia Basin for hundreds of generations. In the 200 years since the Lewis and Clark expedition, however, the River and its tributaries have been dramatically altered, bringing many benefits for humans but also some serious costs including the decimation of the once-mighty salmon runs. Water withdrawals, primarily for irrigation, are one cause of the River's altered state; while there is sharp disagreement over the impact of existing water withdrawals on Columbia River flows, irrigation is increasingly viewed as a significant part of the problem. While water laws and institutions in the Pacific Northwest now take some account of the needs of salmon, water managers also face growing pressure to authorise new withdrawals that would further deplete flows in the Columbia. This section briefly addresses these background matters, providing some context for a discussion of NAS's Columbia River study.

The Bygone Columbia

In terms of flow, the Columbia is far and away the biggest river in the western United States, with an average annual discharge of 281,000 cubic feet per second (cfs). By comparison, the second-biggest river on the West Coast, the Sacramento, averages 23,000 cfs. The West's most storied river, the Colorado, carries less than haft the water of the Columbia's major tributary, the Snake. The Columbia drains an area of 258,000 square miles, including the vast majority of Idaho, Oregon, and Washington, plus sizable chunks of British Columbia, Montana, and Wyoming. Before it was harnessed by a number of mainstem and tributary dams, the Columbia, with highly variable flows and giant rapids, was a wild river in every sense. The untamed Columbia may have been inhospitable to large-scale hydropower, irrigation, and navigation, but it provided an enormous natural bounty of salmon. The Columbia's annual salmon runs may once have been as great as 16 million fish, a number that is roughly triple the current human population of the U.S. portion of the Columbia Basin.

These mighty salmon runs were the foundation of the Columbia Basin's first human society, that of the native tribes. Human residence in the Columbia Basin apparently dates back at least 12,000 years, and the tribes sustained themselves largely by catching salmon at Celilo Falls and other sites on the Columbia and the Snake. As stated by the Supreme Court a century ago, for these tribes, "the right to resort to the fishing places was not much less necessary than the atmosphere they breathed." Nineteenth century settlers also harvested these runs extensively, and after canning techniques were refined, salmon fishing and canning became one of the region's first major industries.

The Columbia Today, Dams, Water Use, and Salmon

Development of the Basin's water resources, and the resulting alteration of its rivers and decline of its aquatic ecosystems, occurred in three major stages. First, irrigation diversions on Columbia and Snake River tributaries such as the Boise, Yakima, and John Day Rivers began in the mid-to late-1800s. Such diversions were often relatively small in scale but were cumulatively significant, causing many of these rivers to be severely depleted by the early part of the 20th century. Enactment of the 1902 Reclamation Act brought the second stage of development: the construction of major federal dams on Columbia and Snake tributaries to store water for irrigation. Within ten years, the United States Bureau of Reclamation (then called the Reclamation Service) was building major irrigation projects on the Boise, Snake, Umatilla, and Yakima Rivers, and many others would follow. When the Bureau of Reclamation completed a dam on Oregon's Owyhee River in 1932, it was the highest dam in the world at the time. The third stage was the building of major dams on the Columbia and lower Snake River mainstems, largely for hydropower purposes. Construction of the Bonneville and Grand Coulee Dams began in 1933, and within 40 years there were six major federal dams on the Columbia and four on the lower Snake, along with five major nonfederal hydropower dams on the Columbia, an additional five major dams upstream in Canada, and three more nonfederal hydropower dams on the Snake at Hells Canyon. This massive development "flattened" the Columbia River hydrograph, sharply reducing the Columbia's naturally high summer flows and increasing the low winter flows, resulting in flow patterns very different than those in which the Basin's salmon populations evolved.

Today, irrigation is easily the biggest offstream use of water in the Columbia Basin, accounting for at least 80 percent of water withdrawals for human uses. The number-two use, public water supply, is small by comparison but growing along with the region's population. Hydropower generation is the other major water use in the Columbia Basin states, accounting for more than half of the region's electrical supply and nearly two thirds of its generation capacity as of 2004.

The Pacific Northwest received many benefits from development of the Columbia River and its tributaries, but it paid with its salmon runs. Not only did dams and diversions drastically alter flows, they also blocked salmon from reaching vast areas of their historic habitat. When combined with other factors such as overfishing, logging, and other land-use practices, the effects of water development on salmon have been devastating. One leading commentator, Columbia River water expert John Volkman, has summarised the decline as follows:

The runs were still relatively large in the 1950s, but with the closing of the floodgates at the last Snake River dam, accelerated timber harvest and all the other changes, Snake River salmon populations went into serious decline. The declines became "synchronous" and widespread in the late 1960s. From historic peaks ranging from ten to sixteen million adult fish, the Columbia runs declined to something like a million. As bleak as this number is, it understates the decline of the wild salmon stocks that scientists see as the "seed corn" for the salmon run, only a handful of populations, the Hanford Reach fall chinook, the Wenatchee River sockeye and several summer steelhead stocks in Oregon's John Day River, for example, are still considered healthy.

Many populations of salmon and other anadromous fish have been listed as threatened or endangered under the Endangered Species Act (ESA). Efforts to save and restore Columbia Basin salmon populations have included some measures aimed at ensuring adequate flows in the Snake and Columbia Rivers. For example, when the Northwest Power Planning Council a regional entity charged with seeking appropriate balance between hydropower generation and ecosystem protection in the management of federal hydropower dams in the Basin developed its original plan for the Basin's fish and wildlife in the 1980s, the key element was a "water budget" to provide mainstem flows for salmon. More recently, the Bureau of Reclamation has essentially rented water from Idaho irrigators in an effort to provide an additional 427,000 acre-feet of Snake River water for salmon. Perhaps most significantly, federal agencies operating under the ESA have established flow targets on the Columbia and Snake Rivers for salmon. But Columbia River flows commonly fall below these targets, especially in the summer, largely because of water withdrawals for irrigation and other purposes. A federal study in the mid-1990s found that in the driest summers, average Columbia River flows at McNary Dam fall 90,000 cfs short of the target levels for salmon, primarily because of upstream water withdrawals for irrigation.

Water Allocation by States Under Prior Appropriation

In the Columbia Basin, as in the rest of the United States, water allocation is chiefly a state matter, and water use is regulated primarily under state law. States have occupied the field of managing water resources largely because the federal government has often deferred to the states in this area, allowing each state to establish its own rules for how its waters will be allocated and used. Federal agencies also play major roles in the management of Columbia Basin waters, especially where dam operations are concerned. For example, the United States Army Corps of Engineers operates major dams on the Columbia, the Snake, and several of their tributaries, largely for hydropower and flood control; the Bureau of Reclamation operates dozens of dams throughout the Basin, primarily for irrigation; and the Federal Energy Regulatory Commission licenses more than a dozen nonfederal hydropower dams in the Basin, including major facilities on the Columbia and Snake mainstems. Thus, Congress has left the states in charge of allocating water among users and regulating their use, even though the federal government has been largely responsible for developing and managing the Basin's water resources, and even though these resources are shared by two nations and several U.S. states.

Washington, like Idaho, Montana, and Oregon, allocates water under a statutory system based on the prior appropriation doctrine. For purposes of this Chapter, several classic features of prior appropriation are worth noting. First, water is a public resource subject to the control of each state. Second, states recognise private rights to use this public resource if the water is diverted from its natural course and applied to a "beneficial use" such as irrigation, industry, or domestic water supply. Third, such "water rights" generally last forever. Fourth, each water right's priority is based on its seniority, so that in times of shortage, older ("senior") water rights get all the water they are entitled to before newer ("junior") rights get any. Fifth, each water right has a specific purpose of use, place of use, and point of diversion; for example, a particular water right may authorise irrigation of a specifically described 200 acres in Benton County, Washington, with water to be taken from a specifically described point on the Columbia. Sixth, a water right may be changed (or "transferred") temporarily or permanently to a new place of use, type of use, or point of diversion, but only if the state approves the change based on a finding that it would not harm other water users. Thus, an irrigator may sell or lease her water rights to another irrigator, or to a city or some other water user, but such changes may not adversely affect other users on the same system, nor can they improve a water right's priority date or increase the maximum quantity of water to which the user is entitled.

The Columbia Basin states now require that any person seeking to make a new use of water (rather than just continue an ongoing use) first obtain a permit from the relevant state agency. While each state has its own requirements, in general a state agency will not issue a new water use permit if it finds that the available water supply has been fully allocated to existing users or if the permit might somehow violate the public interest. A permit provides no guarantee of a reliable water supply, however, because of the prior appropriation seniority system a permittee cannot use water if all the available supply is required to satisfy older, higher-priority rights.

Traditionally, prior appropriation did not protect water flowing in its natural course ("instream flows") primarily because the doctrine required a diversion of water from its natural course as an indispensable element of a water right, but also because of doubts about whether instream uses were truly "beneficial." Led by Oregon and Washington, the western states began easing these restrictions in the latter half of the 20th century, and today all the Columbia Basin states recognise water rights for instream flows. Instream flow protections are established differently from traditional water rights under prior appropriation because they generally are created only in response to a request from a state agency, primarily for the purpose of protecting sufficient flows to provide adequate habitat for fish populations. Once created, however, instream flow protections are much like regular water rights in that they carry a specific location and quantity of water, along with a fixed priority date that determines their place in the priority system.

Washington established instream flow protections for the Columbia River mainstem in 1980. Users with water rights senior to 1980 could continue to use water if the Columbia fell below these specified levels. Ecology continued to issue new permits, however, and these post-1980 users could have their water use cut off if the River fell below the minimum flows; these junior rights came to be known as "interruptible." In the critical drought year of 2001, however, Ecology allowed irrigators holding these interruptible rights to continue their use despite low flow conditions on the Columbia.

Through the 1980s and into the 1990s, the states continued to issue permits for water withdrawals from the Columbia and Snake as the Basin's salmon populations continued to decline. In response to ESA listings of Columbia Basin salmon populations, Idaho, Oregon, and Washington all imposed moratoria on the issuance of further water withdrawal permits from the Columbia and Snake Rivers. Although the Washington and Oregon moratoria did not absolutely prohibit all new diversions, they did represent an effort to prevent additional water withdrawals from causing further risks to salmon.

Pressure for Expanded Uses (New Rights and Uninterruptible Status)

Washington's moratorium was not popular with agricultural interests and others especially in the eastern half of the state who believed that the economic importance of new Columbia River water withdrawals outweighed any risks to salmon caused by reduced flows. In 1997, the Washington legislature passed a bill to lift the moratorium and new Governor Gary Locke signed it, although he promised that Ecology would not "process any pending or future applications for new appropriations until instream flows are established and a determination is made that water is available for withdrawal or make decisions for new appropriations that would jeopardize regional salmon recovery efforts." Thus, lifting the moratorium did not immediately result in the issuance of new permits, but it raised the pressure on Ecology to authorise new withdrawals.

Some existing Columbia River water users were also asking Ecology to improve the status of their water rights. Specifically, those users whose water rights were junior to the 1980 instream flows on the Columbia and therefore "interruptible" during low-flow years were seeking to have their rights converted to "uninterruptible" status. Ecology had issued interruptible water rights exceeding 172,000 acre-feet, and these users were essentially seeking to have their water rights elevated in priority over the 1980 instream flows. Serious questions persisted, however, about the potential for new or expanded water withdrawals to harm salmon populations by depleting summer flows in the Columbia River mainstem.

The National Marine Fisheries Service, responding to a proposal for a large new irrigation withdrawal from the Columbia River in Oregon, found that the new withdrawal could jeopardize salmon by further reducing mainstem flows that already fall below target levels in most years. In Washington, two environmental organisations petitioned Ecology in 1999 for a new moratorium on Columbia River water withdrawals until it established instream flow needs for salmon.

Under pressure from all sides regarding Columbia River water, Ecology launched its "Columbia River Initiative," with the stated purpose of allowing access to new water withdrawals while providing support for salmon recovery. Ecology acknowledged that decisions on Columbia River water management had become increasingly controversial, and that "lack of scientific consensus [had] resulted in gridlock." In the hopes of resolving the gridlock, or at least providing a scientific foundation for a defensible water management regime, Ecology asked the National Academy of Sciences for a study of the relevant issues on the Columbia.

The National Academy of Sciences Study

NAS and the Committee

NAS, which conducted the Columbia River study, is the most prestigious national entity for scientific research. As NAS stated in one of its reports, "NAS and its associated organisations constitute what has been called 'The Supreme Court of Science' and the most important independent expert advisory mechanism for science and technology in the United States and perhaps in the world." The National Research Council (NRC) is the working arm of NAS, carrying out most of the studies done in its name. NRC fulfills this mission primarily by "calling upon a wide cross section of the nation's leading scientists, engineers, and other professionals, who serve on its committees without pay."

The NAS study process has multiple safeguards to provide studies that are balanced, objective, and well-grounded in science. Once a topic is approved for study, NAS seeks to ensure that each committee has the appropriate balance and expertise for its assigned task. All reports are subject to peer review for scientific validity and to further review to ensure that the committee's conclusions and recommendations are supported by the body of its report. NAS drew its committee for the Columbia River study from a variety of disciplines and backgrounds. The Chair, Ernest T. Smerdon, was an expert on water resources and agricultural development who recently retired as Vice Provost and Dean of the College of Engineering and Mines at the University of Arizona.

The other 12 committee members included a professor of agricultural and resource economics at Oregon State University; a consulting fisheries biologist from Eagle, Idaho; the director of the International School for Water Resources at Colorado State University; a consulting fisheries scientist from Redmond, Washington; a professor of social ecology with water resources expertise from the University of California-Irvine; a professor of ecology from South Dakota State University; an emeritus professor of zoology and limnology from the University of Wisconsin-Madison; a water quality expert retired from the United States Geological Survey; an engineering professor from the University of Colorado; a watershed management coordinator from the Michigan Department of Natural Resources; an emeritus professor of hydrologic science from the University of California-

Davis; and an attorney with expertise in water rights and resources. The committee staff was led by Study Director Jeffrey W. Jacobs from NRC's Water Science and Technology Board. The NAS committee held a series of public meetings, three in Washington State and the last in Washington, D.C., where it heard from Ecology staff and a variety of "others with specific interests or expertise." A variety of experts from government, major universities, and the private sector reviewed the draft report. The "prepublication" final copy of the report was released to the public on March 31, 2004.

The Committee's Charge: Salmon Science and Water Management Scenarios

Ecology's requests to NAS were both general and specific. The general request was for a review of the science underlying decisions regarding new water withdrawals from the Columbia River and the potential effects of such withdrawals on salmon. The Report describes the general portion of the committee's charge as follows:

A key issue in this study was the pending applications for additional water rights permits from the mainstem Columbia River in the State of Washington, applications which have been on hold for some time. Our committee's charge was to consider the implications for [sic] potential additional withdrawals for Columbia River salmon and to comment on the body of scientific knowledge related to this issue and its implications. The committee was not charged to review all ecological issues (of which there are many) across the basin which affect salmon but rather to conduct a more focused investigation regarding conditions in a stretch of the mainstem Columbia River in the State of Washington. Nor was the committee charged with recommending policy decisions but rather was requested to review the scientific information available by decision makers and to comment on it.

More specifically, Ecology presented five management scenarios involving proposals for water withdrawals from the Columbia River mainstem and asked NAS to respond to these scenarios. The Report notes that these five scenarios "contained many assumptions and actions related to water withdrawal quantities, management actions, and water use fees.. Several possible management actions did not contain enough specificity to enable detailed evaluation." In general, Ecology's management scenarios (except for the highly uncertain fifth scenario) would involve some combination of the following: i) issuing new permits for water withdrawal from the Columbia up to some defined quantity, ii) converting existing "interruptible" water rights to "uninterruptible" status so that use of these converted rights during low-flow periods would not be curtailed to protect flows for salmon, iii) requiring "state of the art" water use efficiency practices for new or converted water rights, iv) imposing annual water use fees applicable to new or converted water rights, v) establishing new mechanisms for permanent or temporary transfers of existing water rights into new uses, and vi) various other management measures. Ecology's five scenarios may be summarised as follows:

Scenario 1: Ecology would:

- issue new permits from the Columbia up to 1 million acre-feet (AF) per year;
- convert existing interruptible water rights to uninterruptible status;
- require "state of the art water use efficiency practices" by users of all new and converted water rights;
- reassess the management program for Columbia River water withdrawals periodically, with formal reviews at ten years and twenty years, to reflect the latest scientific information;
- establish a functioning water market or water bank for Columbia River water rights;
- assume that Oregon may withdraw up to an additional 600,000 AF from the Columbia mainstem for its future needs; and
- protect an additional 427,000 AF of Snake River instream flows through the Washington portions of the Snake and Columbia Rivers.

Scenario 2: Same as Scenario 1, except that Ecology would

- impose an annual fee of $10 per AF on all new and converted water rights;
- use proceeds of these fees to acquire water rights to improve streamflows, restore salmon habitat in the Columbia and its tributaries, and possibly develop new water storage projects;
- issue new permits from the Columbia up to 700,000 AF per year; and
- issue additional new permits up to 300,000 AF per year only upon a showing that "conservation investments were in place for a majority" of existing Columbia River water users.

Scenario 3: Same as Scenario 2, except that Ecology would:

- charge $20 per AF annually (instead of $10) for new and converted water rights; and
- provide financial support for new water conservation measures and potential water supply improvement measures, such as new storage projects.

Scenario 4: Ecology would:

- require "direct mitigation in the mainstem" for any new water use on the Columbia, i.e., require that new withdrawals be fully offset by acquiring existing water rights, implementing new water conservation projects, or using new storage;
- impose a fee of $30 per AF annually on converted water rights; and
- use the proceeds of the $30 annual fee to acquire water rights for instream flows or fund other habitat improvements in the Columbia mainstem and tributaries.

Scenario 5: Ecology would consult with state, tribal, and federal fisheries managers on each proposal for an individual new water right. "Under this scenario whether or not mitigation is required and the type and quantity of that mitigation are decisions made on each permit on a case-by-case base [sic] as a result of the consultation."

It is worth noting a couple of issues that Ecology did not ask NAS to address. First, there was no request to determine specific minimum flows of the Columbia River for salmon survival and recovery that is, NAS was not asked to quantify how much water salmon "need" in the Columbia mainstem. Second, Ecology did not ask for an assessment of the harm to salmon (if any) caused by existing water withdrawals; instead, NAS was essentially directed to take existing withdrawals as a given in determining the impacts of potential new withdrawals.

In reviewing the science regarding the impacts of Columbia River water management on salmon populations, NAS addressed a number of subjects. NAS examined trends in water flows, temperature, and quality in the Columbia River; the relationship between these parameters and salmon survival; the existing use and allocation of water from the Columbia; the laws and institutions that govern water management in the Columbia Basin; and factors (such as ocean conditions and climate change) that create uncertainty for the future because they are hard to predict and control. NAS devoted the great majority of its Report a full 160 pages to an examination of existing scientific data and other information on these subjects. This portion of the Report is summarised below in the briefest possible terms.

In its chapter on "Hydrology and Water Management," the Report notes that human activities particularly the construction of 50 plus major dams on the Columbia River and its tributaries have dramatically altered the River's natural flow patterns. Much of this change, though certainly not all, has been caused by the operation of the Federal Columbia River Power System the system of large hydropower dams on the Columbia and Snake River mainstems.

The Columbia also has a number of major water withdrawals for offstream uses, primarily for the purpose of irrigation. These existing withdrawals have little impact on Columbia mainstem flows during the winter months, but in July and August they may divert nearly ten percent of the flow in average years and more than fifteen percent in very dry years; the report stresses the need to consider the effects of water withdrawals during these low-flow periods. The report also notes that mainstem Columbia water temperatures have been steadily increasing over the years, creating an additional risk to salmon, and that climate change could exacerbate this problem in the future. This chapter concludes by saying, "Although precise cause-and-effect mechanisms are hard to define clearly, the changes in Columbia River hydrology identified in this chapter have greatly affected the basin's salmon populations."

The following "Environmental Influences on Salmon," acknowledges that Columbia Basin salmon are among the world's most intensively studied fish species, but that much is not yet known about them. After noting that Columbia Basin salmon populations have declined precipitously over the past several decades, the Report examines research, modelling, and alternative hypotheses regarding the declines, focusing on studies regarding the effects of Columbia River flow levels and velocities on salmon survival. In the end, NAS was unable to reach a general conclusion on the importance of these factors, but found that critically low flows and high water temperatures would present a significant problem:

Within the body of scientific literature reviewed as part of this study, the relative importance of various environmental variables on smolt survival is not clearly established. When river flows become critically low or water temperatures excessively high, however, pronounced changes in salmon migratory behaviour and lower survival rates are expected.

The Report then examines "Water Laws and Institutions," highlighting the many federal, tribal, and state governmental entities that have a hand in Columbia Basin water management. It also notes that a multistate entity, the Northwest Power and Conservation Council (formerly the Northwest Power Planning Council), has been authorised by Congress to balance power production with fish and wildlife conservation in the Columbia Basin. It goes on to discuss tribal claims to the Columbia Basin's water and salmon resources, and identifies a few of the more important federal rights and obligations regarding the Columbia (including the protection of salmon and other endangered species). The chapter ends with a discussion of state water laws and institutions, with a primary emphasis on Washington. Idaho, Montana, Oregon, and Washington all allocate rights to use water under the prior appropriation doctrine and require anyone seeking to make a new use of water first to obtain a permit from the state. The next "Better Management of Existing Water Supplies," essentially examines the prospects for meeting water demands through reallocation of existing water supplies rather than through new permits. The chapter begins with a discussion of studies on the economic value of water in various uses such as agriculture, municipal water supply, hydropower generation, and recreation. The Report finds substantial differences in the dollar value of water based on the category of use and explains that "[t]he significance of the differences across uses is that there is a great potential to promote economic growth and increase overall social benefits by transferring increments of water between uses (from low- to higher-value uses)." It then discusses how water markets and water banks may facilitate such transfers on either a temporary or permanent basis or notes that water banks and markets are growing in popularity across the West. The chapter then notes that water conservation measures may provide a source of "new" water supply, but they tend to be expensive often too much so for many agricultural users. The chapter concludes by urging Ecology and other water management entities to promote water marketing and conservation as viable means of providing water for economic growth without the need for additional withdrawals that could harm salmon.

Water supplies in the Basin are uncertain, both from year to year (drought) and over the long term (climate change). Water demands are also uncertain, especially because there are many entities with claims or interests in Columbia Basin water but no single entity with overall management authority.

Risks to salmon grow with Columbia Basin water demands, and they are greatest when high demands coincide with naturally low flows in the River. The take-home message is that water right permitting decisions should account for major uncertainties regarding future water supplies and demands, especially given all the other factors that pose serious risks for salmon populations. As the Report states, "given the current [legal/institutional] setting and likely future climatic and other trends, additional water withdrawals from the river during seasons characterised by low flows (particularly in drought years) will pose additional risks to salmon survival," which Ecology should consider in making decisions about proposed withdrawals during such low-flow periods.

The Report goes on to address how Ecology and other Columbia Basin water entities should approach such decisions on water permitting and management.

First, it notes that an "interjurisdictional water organisation" for the Columbia Basin involving the Canadian and U.S. governments, state and provincial governments, and tribes could reduce uncertainties and risks arising from uncoordinated decision making by each of these sovereign entities.

Second, the Report suggests dealing with multiple uncertainties through an adaptive management approach whereby today's incremental management decisions not only promote the development of new information, but also preserve maximum flexibility to make future changes in response to such information.

Third, and perhaps most interestingly, the Report states that science can inform policy decisions on matters such as water permitting, but cannot make these decisions or eliminate all uncertainty surrounding them. "Progress toward 'comprehensive' management cannot be accomplished through scientific inquiry alone, but rather requires stakeholders and management agencies to work with scientists in a collaborative learning process, such as that framed by adaptive management principles."

Consistent with Ecology's request, the Report effectively offers two sets of findings and recommendations: one primarily addressing general concepts, the other reacting more specifically to Ecology's five management scenarios as described above.

Six General Findings and Recommendations

A brief "epilogue" chapter begins by noting that various human activities have affected Columbia River salmon populations and that the long-term decline of these populations is undeniable. It then identifies "six key findings and recommendations" of the Report, which may be summarised as follows:

1) Critically low flows or high water temperatures can be expected to cause significant changes in salmon migratory behaviour and lower salmon survival rates.
2) Ecology and its Columbia Basin counterparts should continue to study and consider using water transfers and market-based programs in lieu of issuing new water right permits.
3) Converting water rights to uninterruptible status is not recommended because it would reduce management flexibility in periods of low flows and high water temperatures.
4) Thorough, inclusive management strategies require not only sound science, but also readiness on the part of decision makers to act despite uncertainty.
5) Decisions on new water permits are ultimately matters of policy; but because new withdrawals would increase risks to salmon in low flow/high temperature periods, any new permits should provide Ecology with the authority to halt withdrawals during such periods.
6) Washington and other Columbia Basin jurisdictions should establish a joint forum to document and discuss the effects of significant proposed new water withdrawals.

Reaction to Ecology's Management Scenarios

The Report also offers a few paragraphs of "evaluation and commentary" regarding the five potential water management scenarios identified by Ecology. In these paragraphs, NAS addresses a number of significant water policy issues in language that is clear, concise, and direct. Because the next section of this Chapter focuses heavily on the water policy implications of this piece of the Report, much of it appears here verbatim: In general, the adoption of concepts related to adaptive management, such as periodic review and adjustment of the program and monitoring, and market-based conservation strategies such as conservation, the use of water markets (or “banks”), and charging for water rights, is commendable.

A pervasive aspect of the scenarios is the lack of a comprehensive basin-wide assessment of water uses and needs as a context for evaluating permit applications. Small (relative to the flows of the Columbia River) withdrawal and permitted volumes will have only small, if not miniscule, effects on the water budget of the basin as a whole. All water uses accumulate, however, both in Washington and elsewhere along the mainstem, as well as the along [sic] tributary streams. If future demands for water increase (which seems likely given recent and projected demographic and economic trends), the accumulation of risks to salmon survival will be all the greater (given the variety of risks that affect salmon survival, assigning precise and credible levels of risk to changes in flows and temperature is extremely difficult). These effects would be magnified by reductions in low flow that could attend prospective climate warming as well as during periodic unfavourable ocean conditions. The lack of a comprehensive basin-wide management structure hampers the ability to make comprehensive judgments (both in time and over space), and it supports this report’s recommendation for creating a basin-wide framework for coordinating water use data and strategies.

- Conversion of interruptible to uninterruptible water rights (Scenarios 1-4). Conversion of interruptible water rights to uninterruptible status makes adaptive responses more difficult. Interruptible water rights are interruptible so that at times of scarcity, instream flows can be protected. Making any out-of-stream right uninterruptible reduces flexibility to retain water in the river when salmon need it most during periods of high demand and low flows.
- Re-evaluation at 10 and 20 years (Scenarios 1-3). The idea of re-evaluating the scenarios periodically is excellent. For this re-evaluation to be meaningful, however, the program needs to be designed so that any aspect of it could be undone (reversed) if the evaluation calls for such a reversal. No evidence is provided of any such reversibility. Instead, the result will be decreasing reversibility by allowing for some interruptible water rights to become uninterruptible. In some cases, more frequent re-evaluations might be necessary.

- Charge for water rights (Scenarios 2-4) Charges for water rights in this scenario appear to be arbitrarily selected and out of proportion to the probable costs of mitigation and the value of water to the users. For example, the scenarios specify charges of $10 to $30 per acre-foot per year to be used (among other things) to acquire mitigation water in low-flow years. This scenario thus proposes increasing the priority of a water permit for $10 to $30 per acre-foot per year and using the money to buy water for what could be several times that amount.
- Water markets (Scenarios 1-4 and perhaps 5) Water markets, water banks, and other such market-based mechanisms offer potential improvements over existing systems of water allocation. However, restricting markets only to the Columbia River's mainstem, and only to Washington, is narrowly construed. The Department of Ecology already allows for 600,000 acre-feet per year to be used by Oregon in its assumptions, but no allowances are made for uses by Idaho, Montana, or British Columbia, or by tribes.
- Structural storage measures (Scenarios 2-4 and perhaps 5). A lack of specificity in this scenario inhibits the ability to comment extensively upon it. It implies that tributaries are to be used for additional storage (which may have negative consequences for salmon), but the habitat and condition of tributaries are of critical importance for Columbia River water quality and for survival of salmon that use the river. Tributaries should thus be considered for protection and mitigation as well.

Thus, the Report has a great deal to say about water management. It addresses key scientific issues, certainly, but also touches upon law, institutions, and decision-making processes. The Report is also filled with significant water policy implications, not just for Washington, but for the entire Columbia Basin and beyond. The following section identifies and discusses some of these implications.

Water Policy Implications of the NAS Study

The heart of the NAS Report is a scientific finding that may be summarised as follows: Water withdrawals that reduce Columbia River July-August flows will increase risks to salmon, but the magnitude of the risk is hard to estimate because of uncertainty about future water supplies and demands in the Columbia Basin and about other factors that pose risks to salmon. This finding seems rather cautious and unremarkable, but it provides the scientific foundation for the NAS recommendations regarding Columbia River water management. Because these management recommendations flow logically from this scientific basis, they might seem unremarkable too except that they go directly against the western states' long-standing approach to water allocation under prior appropriation, as explained below.

A Joint Forum for Columbia Basin Water Allocation

NAS found that one of the major sources of uncertainty in Columbia River water management was the lack of predictability regarding future water withdrawals, due in part to the lack of basin-wide control or coordination regarding new water permits. Thus, NAS strongly recommended creating a joint forum among Columbia Basin water entities to consider the basin-wide implications of proposed significant new water withdrawals from the Columbia River system. The Report states,

This forum need not entail anything binding other than an obligation to refer the applications. At a minimum, proposed diversions would be subjected to professional and public scrutiny, magnitude of risk, possibilities of mitigation, and system-wide equities. A basin-wide forum for considering withdrawal permit applications would enhance unified water management across the Columbia River basin. The State of Washington and other basin jurisdictions should create a joint forum for documenting and discussing environmental and other consequences of proposed diversions that exceed a specified threshold.

This seemingly modest recommendation for regional review of individual permit applications is contrary to a basic assumption of water management in the West that each state will make its own decisions about allocation of "its" waters to serve its own interests. Of course, states have staunchly resisted any federal government interference with their individual autonomy over water allocation, especially in the context of federal environmental laws. The states have not been so vocal in opposing any potential "interference" by their fellow states, but neither have they commonly banded together in joint efforts to address water allocation issues on interstate rivers. To the contrary, states have routinely fought among themselves over interstate rivers as each state goes all out to protect its individual interests, as shown by the recent battle among states along the Missouri.

Even where states have entered into a compact that purportedly allocates the waters of an interstate river, these compacts have not precluded further litigation among the states over the management and use of the rivers' waters. The chief motivation for some major compacts was the promise of federal water projects that would allow for expanded storage and use of the waters of an interstate river but would be built only if the states could agree on allocation of its waters. It seems fair to say, at least in the West, that states generally have viewed interstate compacts primarily as means of protecting their individual interests. By contrast, NAS has proposed that the Columbia Basin states band together for a fundamentally different purpose: to maintain the ecological commons. In other words, NAS seems to envision a joint Columbia River forum that would priorities protection of adequate flows for salmon habitat rather than allocation of water among the states for maximum "beneficial use." Thus, while state permitting statutes allow rejection of a new water withdrawal that would impair the public interest, the joint forum would essentially ask whether a new permit would harm the public interest of the entire Columbia Basin including the interest in sustaining salmon runs.

The Columbia Basin has an established regional entity, the Northwest Power and Conservation Council (formerly the Northwest Power Planning Council), charged with developing a Columbia River management regime that adequately protects fish and wildlife resources. The Report notes that the joint forum "could be convened within the existing Northwest Power and Conservation Council, which includes broad representation of political entities from across the basin," but stops short of recommending this arrangement. The Council has been criticised for failing to take strong measures to save salmon, especially regarding mainstem flows in the Columbia and Snake. On the other hand, it was the Council which in 1993 called on the states to work together in protecting adequate mainstem flows for salmon, prompting Idaho, Oregon, and Washington to impose moratoria on new Columbia-Snake water withdrawals. The Washington Legislature soon acted to repeal the permitting moratorium, however, which shows the difficulties of convincing any state to forego new water diversions for the good of a broader region even if the other states are willing to participate. Under pressure from the Endangered Species Act, some of the other western states have entered into cooperative agreements regarding water use and endangered species recovery in interstate river basins such as the Upper Colorado and the Platte.

The United States Department of the Interior, a key participant in the Upper Colorado and Platte recovery programs, has taken the position that cooperative regional approaches are the best way to address ESA issues involving river management. Secretary of the Interior Gale Norton has said, "We need more Upper Colorado Endangered Fish Recovery Programs and fewer Klamaths." It remains to be seen whether the Upper Colorado and Platte programs can successfully rebuild populations of ESA-listed species. These two collaborative programs do seem to demonstrate, however, that interstate agreements regarding water and ecosystem management can be workable from the states' standpoint, especially given the alternatives of conflict and litigation.

Conservation and Markets as the First Option for Meeting New Water Demands

The Report recognises that the Columbia Basin will continue to see sizeable new demands for water to supply cities, farms, and other purposes over the next 20 years demands that easily could exceed one million acre-feet in Washington alone and two to three million acre-feet throughout the Basin. The Report does not suggest that these new demands should not receive water, but it does encourage the states to consider ways to satisfy them without simply issuing new permits:

Increased flexibility in managing the Columbia River will require greater emphasis on nontraditional approaches to augmenting water supplies, such as water marketing and water transfers, and greater cooperation of political entities across the basin. These market-based programs may require capital investments in physical infrastructure and human resources investments in experts with skills in fields such as finances, marketing, and public administration. Programs such as water transfers, groundwater banking, and other measures to increase the efficiency of water use hold promise in helping sustain the regional economy in ways that do not require ever-increasing water withdrawals. Although water uses across the basin should not be simply channelled to the highest bidders for water, such measures hold promise for helping support both economic and environmental goals and should be carefully considered.

This recommendation goes against another well-established practice of the western states in allocating water: New water withdrawals should be permitted if there is water legally available for appropriation; and if there is none, the best option is to develop new water supplies through storage projects or pumping groundwater. While many observers believe that the era of major dam building in the West is over, the states have not wanted to concede this point. Water right transfers and other market-based tools have not been the preferred approach to meeting new water demands in the West, as shown by the rather small number of water transactions taking place throughout the region. And the states' approach to water conservation as a management tool is similar to Vice President Cheney's widely quoted view of energy conservation that is, water conservation may be a sign of personal virtue, but it is not a viable water policy. The Report suggests that Columbia Basin states can no longer afford simply to accept all existing water uses as a given or to rely primarily on new permits or new storage projects to meet new water demands. Instead, the states should take steps to encourage greater productivity from existing water supplies that is, to promote more efficient use of water that has already been allocated. For example, the states could subsidize improvements in irrigation systems to reduce evaporation or seepage losses, thus improving their physical efficiency. In addition, the states could help establish water banks or other market mechanisms to facilitate the (temporary or permanent) movement of water from lower to higher valued uses, thus increasing the economic efficiency of water use. Either way, existing water uses could effectively provide a source of supply to meet new demands without the need for additional withdrawals from the Columbia or its tributaries.

By emphasising water conservation and market-based approaches that would encourage (or perhaps even require) changes in established water uses, the Report runs contrary to another fundamental feature of water management in the West: nearly absolute protection of existing uses. In an earlier chapter, I observed that the Columbia Basin states have consistently refused to interfere with established water uses, even where the states have had to bend, change, or ignore traditional prior appropriation hales to do so. Of course, one drawback of protecting the status quo is that it forces prospective new users to look elsewhere for their water supply: "By refusing to apply laws that would reduce the quantity of water legally held by existing users, states limit the amount of water potentially available for new appropriations." To the extent that new demands must be satisfied by shifting water from existing uses, however, the opportunity cost of protecting the status quo goes up.

The Report identifies several areas of the West where market-based approaches are being used to address difficult water resource problems. It cites examples from California, Idaho, Oregon, and Washington, as well as a newly authorised interstate water bank on the lower Colorado involving Arizona, California, and Nevada. The Report certainly could have offered further examples, such as the thriving trade in shares of the Colorado-Big Thompson Project on the northern Colorado Front Range, probably the nation's best-developed water market. In addition, a wide variety of "water banks," many of which have at least some environmental purposes, have been sprouting across the West in recent years. In short, while the NAS recommendation may cut against a strong tradition in the western states, it is entirely consistent with a modern trend toward greater use of market-based mechanisms.

Tougher Requirements for Obtaining New Water Use Permits

Because new water withdrawals would further deplete mainstem flows in the critical July-August period and thus increase risks to salmon, NAS indicated that it did not favour new permits from the Columbia River. At least three of Ecology's proposed water management scenarios did contemplate new withdrawals, however, so the Report addresses permitting issues in some detail. One of the Report's more significant comments on permitting is the following:

A key problem in managing the basin's water is that water permitting decisions are currently made in a piecemeal fashion, with little to no consideration of their effects on other users or their degree of consistency with other decisions across the basin. If water resources and risks to salmon survival are to be better managed, Columbia River water permitting decisions must be made in a more holistic fashion, with consideration of how additional diversions would affect other users and sectors across the entire river basin.

While this passage deals in part with the need for regional decision making on new water withdrawals, it also makes a more fundamental point: Permitting decisions should consider the full range of effects that could result from new water withdrawals, including environmental impacts that may be felt both locally and regionally. Here again, this recommendation is inconsistent with the western states' traditional approach to water allocation. Much of western water law rests on a fundamental policy of maximum utilisation of scarce water resources for human "beneficial uses." States generally have followed this policy in making decisions on new water withdrawals, commonly issuing a requested permit unless no unappropriated water is deemed to be available for the new use. In addition, states have been reluctant to stop issuing new permits despite great uncertainty about whether the water resource can sustain the authorised uses. Washington's 1997 legislation lifting the moratorium on new Columbia River water withdrawals, enacted despite ongoing concerns and uncertainty about flow requirements for salmon, exemplifies this "damn the torpedoes" approach to permitting.

Even though the great majority of western states (including those in the Columbia Basin) have statutes requiring that the "public interest" be considered in permitting decisions, the Report's recommendations on this topic nonetheless seem to call for significant change in state practices. First, while state laws call for consideration of public interest factors, it is not at all clear that the states actually give these factors much weight in deciding on permit applications. While it is hard to determine the actual importance that state agencies give public interest factors in permitting, it seems significant that few if any reported judicial decisions involve a state agency's denial of a permit (or even issuance of a permit with stringent conditions) in the exercise of its power to protect the public interest. Second, assuming that state agencies do seriously consider the public interest in deciding on permits, existing laws probably limit each agency to weighing public interest factors relevant to its own state; Idaho's permitting statute is even narrower than that, providing for consideration only of the "local public interest." By contrast, the Report recommends that agencies take a broader view of the public interest in permitting, considering the implications of new water withdrawals for the entire Columbia Basin.

Instream flow programs, whereby states establish minimum flows in specific river reaches, represent an additional means of safeguarding fish habitat from the impacts of new water withdrawals. The Columbia Basin states establish minimum instream flows for protection of fish habitat, and in some areas these minimum flows have claimed the last available water in a river, effectively precluding any new withdrawals. Thus, state instream flow programs can provide meaningful protection against new depletions, but this protection is limited to those streams where instream flows have been legally created and to the specific quantities of water set aside. By calling for full consideration of the risks associated with new depletions, however, the Report suggests that these existing state programs may not go far enough in protecting fish habitat. Under NAS's recommended approach, state permitting agencies should recognise the impacts of new depletions even on streams that do not have a legally established instream flow. Moreover, the agencies should recognise that existing minimum flow levels may be inadequate, and new depletions may pose important ecological risks even if they do not reduce flows below these existing levels. Thus, the Report implies that states may need to go above and beyond their instream flow programs in order to protect fish habitat from impacts caused by new permits.

Finally, the Report discourages the states from relying on "mitigation" measures to address the impacts of new water withdrawals on salmon. NAS specifically criticizes the idea of allowing new water withdrawals that would further deplete Columbia River flows in exchange for an annual payment of $10 to $30 per acre-foot, saying that these payments would likely cover only a fraction of the probable cost of acquiring replacement water. More generally, the Report questions the effectiveness of mitigation measures, especially those that are not onsite and in kind, as a strategy for addressing the potential harms of new water withdrawals:

Although the idea of "mitigating" impacts is attractive, the reality of most mitigation measures is that they are not well coordinated; that is, a management agency may attempt to offset harmful impacts of water withdrawals in one part of a river system with mitigation measures (e.g., ecosystem restoration) elsewhere. The ultimate outcomes of such varying actions, however, are difficult to accurately predict, measure, and compare (if indeed they are ever measured and meaningfully compared, which they often are not), thus making it difficult to determine if "mitigation" was actually achieved.

In the end, the Report acknowledges that permitting decisions are matters of public policy, and that new permits may be issued for Columbia River water withdrawals. In making permitting decisions, however, the states should consider all the ecological risks and other impacts that may result from new water withdrawals, and ensure that mitigation measures effectively address those impacts. Such requirements would make new permits less certain and more expensive to obtain, leaving them a less attractive option for prospective water users. Of course, this result is fully consistent with the Report's suggestion that water conservation and markets, rather than new permits, should be the first option for satisfying new water demands in the Columbia Basin.

Greater Flexibility in Water Allocation and Management

The Report emphasizes that Columbia Basin water managers face enormous uncertainty regarding future water supplies and demands, while salmon populations (and those responsible for managing them) face even greater uncertainty. It recommends that water managers respond to these uncertainties by employing an adaptive management strategy that is, by taking incremental steps to generate new information and by maintaining the flexibility to make changes in response to such new information. The Report concludes, however, that state water allocation and management under prior appropriation has been too rigid, and that better flexibility in these areas is crucial:

Water withdrawal applications and permitting decisions are highly contentious in both the State of Washington and other parts of the Columbia River basin. Inflexibilities in traditional western U.S. prior appropriation doctrine have contributed to these tensions. A greater degree of flexibility in traditional water permitting and rights processes is paramount to better water management and to decreasing tensions and conflicts in the basin. This report recommends implementation of a joint basin-wide water management forum and the pursuit of nontraditional water marketing and conservation strategies. A water permitting and rights process that more explicitly recognises seasonality of flows should also be devised. Decisions regarding the granting of new water rights are issues of public policy, but additional water withdrawals during the critical high demand and low-flow periods discussed in this report will increase the risks of survival to listed salmon stocks. It will also decrease the flexibility of management institutions to allocate water between different uses in critical low-flow conditions. To increase the flexibility of water management organisations and programs and to better recognise uncertainties regarding future supplies and demands, a new permitting process should be created that allows for withdrawals to be discontinued during periods of low flow and periods of comparatively high water temperature.

The kind of flexibility the Report recommends is the exact opposite of what western water laws and water allocations seek to provide maximum certainty for water users. Several basic elements of western water rights offer a high degree of security for water users: Water rights are private property rights (even though water is a public resource by law), they last forever provided they are regularly used, and they hardly ever contain any sort of "reopener" provision allowing the states to impose significant new restrictions for the purpose of protecting public values. Although the bedrock principle of "beneficial use without waste" could be applied to cut back excessive water uses under existing rights, a state's invoking this principle would create potential uncertainty for users and that is surely a major reason why states have rarely used this authority. Water right adjudications, proceedings which attempt to determine all existing water rights in a river basin, are unbelievably long and complex proceedings, but many western states are now pursuing them in an effort to ensure greater long-term certainty for their water users. Secretary of the Interior Gale Norton has emphasised that western water law (and water users) value certainty above all else:

If we yield to this temptation [to resolve water disputes based on politics or public opinion rather than established water law], we will destroy the one thing that everyone who needs water seeks the most certainty and predictability about how it will be allocated in times of shortage. Without certainty, people cannot plan for their future. Without certainty, it will be far more difficult for market-based transfers to succeed. Without certainty, we will not be able to protect our environment. It is at these times that we must renew our commitment to resolving disputes and answering tough questions within the framework of enduring principles of law.

Thus, while NAS clearly recognises that its call for flexibility would represent a significant change in the western states' traditional approach to water allocation, in reality the Report calls for a truly fundamental shift. The management scenarios proposed by Ecology would provide very little of the kind of flexibility recommended by NAS. These scenarios emphasize 1) the conversion of interruptible water fights to uninterruptible status and 2) the approval of permits for new water withdrawals, with certain provisions for water conservation and mitigation. The scenarios make no provision for limiting these allocations to a certain period of years or for revising them at a later time based on new information or changed conditions. The scenarios did propose that Ecology revisit its Columbia River water management program in ten years and twenty years; NAS endorsed this idea, but noted that such a re-evaluation would be meaningful only if any aspect of the program could be reversed based on the re-evaluation. "No evidence is provided of any such reversibility. Instead, the result will be decreasing reversibility by allowing for some interruptible water rights to become uninterruptible." Thus, NAS correctly saw a basic conflict between maintaining flexibility to make changes needed to protect salmon, and providing certainty to water users through permanent, uninterruptible water rights.

The simple premise of the NAS recommendations is that salmon should not bear all the risk associated with new water withdrawals in the Columbia Basin. Thus, the Report urges states to avoid permanent, fixed water allocations in favour of an adaptive management strategy: "Decisions and policies should promote flexibility while their outcomes are being evaluated and better understood." NAS observed that water banks, water markets, and incentives for water conservation all "tend to support greater water management flexibility and merit careful consideration." If new permits are issued, the Report recommends preserving flexibility by including specific conditions to allow withdrawals to be discontinued during critical periods. The NAS Report did not mention some other permit conditions that could improve flexibility, such as reopener clauses (allowing a change in permit terms based on some future contingency) or time limitations (requiring the applicant to apply for renewal after a certain number of years).

Some people might read this NAS recommendation as giving certainty to salmon and shifting all the risk to water users, but that interpretation seems unrealistic. In theory, adaptive management has been a key principle of Columbia Basin salmon recovery efforts since 1984. In practice, the agencies overseeing these efforts have generally been quite cautious, avoiding initiatives that would fundamentally change resource allocation and use to the detriment of key industries. If state agencies commit to the principle of adaptive management of Columbia Basin water resources a big "if"they could implement it only by taking actions opposed by water users and their allies. Water use interests remain powerfully influential in every western state, as demonstrated by the 1997 Washington statute lifting the Columbia River permitting moratorium. Adaptive management will work in this context only if politics allows it to work, and that, too, is a big "if."

With its recommendations for a joint Columbia Basin water forum, water markets and conservation as the first option to supply water for new uses, tougher requirements to obtain new permits, and greater flexibility in water allocation and management, the Report clearly has major water policy implications. It remains to be seen, however, whether the Report will actually bring about water policy changes in Washington or anywhere else. The concluding section of this Chapter addresses this question.

One of the most notable features of water policy in the West is its resistance to major changes. Because NAS recommends significant changes in the western states' usual approach to water allocation and management the kind of changes that many states have at best been slow to adopt one must ask whether the Report is likely to have any significant water policy impact.

The Report is most directly relevant to Washington, of course, the state that requested the NAS study. Because the Report was not released until March 31, 2004, it was too early as of this writing to determine how it might affect Washington's management of the Columbia. On the other hand, the reaction of key stakeholder groups has been both unmistakable and unsurprising: environmental groups love it, while agricultural interests hate it. One irrigation group told state officials that their characterisation of the Report's findings regarding the risk to salmon from new July-August water withdrawals "reflects a fanatical desire to support a 'no net loss' water policy, an extraordinarily ill-founded basis for your 'water bank' proposal." The agricultural community's harsh criticism of the Report shows how difficult it is to resolve contentious water issues through science, or even to gain acceptance of a National Academy of Sciences study by those who oppose its conclusions.

Ecology's initial response to the Report came in the form of a discussion proposal unveiled on April 30, 2004. Ecology identified two major goals, "meet the water supply needs" and "reduce risks to fish," requiring 500,000 to 750,000 acre-feet per year in the period from

April through September; one-third of this water would be allocated instream for salmon, and two-thirds offstream for human consumptive uses. To provide this water, Ecology proposed a state-sponsored water bank, with water made available from conservation measures, storage projects (existing and new), and water right acquisitions. Funding would come from a mix of state and private sources, including fees for new water users. Six months later, Ecology produced both a draft administrative rule and a draft bill regarding Columbia River water management. These draft documents, like the initial proposal, call for Ecology to acquire water for the Columbia mainstem from a variety of sources, to allocate two thirds of this water for consumptive uses and one third for instream flows, and to use both public and private money to fund these acquisitions. Thus, the agency proposed a program that is consistent with the NAS recommendations favouring conservation and markets over new permits, although the proposal ignores the recommendation for regional review of permits and is somewhat contrary to NAS's call for greater flexibility in water allocation and management. As of early 2005, Ecology's plans were to accept comments on the proposal through May 12 and adopt a final rule in June 2005.

Will the Report directly affect water policy in states beyond Washington? That seems unlikely, at least for now. The Report focuses specifically on the Columbia, especially the Columbia mainstem, so its findings and recommendations are most relevant for Idaho and Oregon, the two states (other than Washington) that lie predominantly within the Columbia Basin. These states have already established rules that seriously restrict the issuance of new permits in the Columbia River system, so they may believe that their policies are already consistent with the Report. Moreover, there has been no significant push to establish the kind of joint Columbia River water forum recommended in the Report. This situation could change if one of the Columbia Basin states begins to authorise major new water withdrawals; such action would cause would-be water users elsewhere in the Basin to turn up the pressure on their own state governments to ease restrictions on new water allocations from the Columbia and its tributaries. Thus, if one state begins to issue new Columbia permits (despite the opposition of environmental groups and other salmon advocates), others may well follow suit. At that point each state could be expected to face questions on whether its water policies were truly consistent with the NAS recommendations, and the proposal for a regional water forum may seem more ripe for discussion. Thus, if the floodgates or more aptly, the headgates open on major new Columbia River water withdrawals, the Report and its recommendations may gain added significance throughout the Pacific Northwest.

Western states outside the Columbia Basin might dismiss the Report as irrelevant; every river system is unique and every species has its own biological requirements, so the science of water and salmon in the Columbia does not translate directly to other places. Moreover, most western states might disregard any study of the Columbia because they see it as a bigger river with bigger ESA issues than may be found in the more and parts of the West. While the Columbia certainly has more water than other western rivers, it shares many of their common characteristics: Its natural flow patterns have been dramatically altered by human development; its waters have been extensively allocated for offstream uses, primarily irrigation; its management is shared by numerous state, federal, and tribal entities; it provides habitat for fish and wildlife species protected by the ESA; and it faces new demands for water for municipal, agricultural, and other uses.

In these respects the Columbia is a classic western river, only bigger. And given NAS's finding that new water withdrawals would raise significant ecological concerns on the mighty Columbia, it might well identify equal or greater concerns on smaller rivers that provide habitat for rare fish or wildlife species if, of course, it was asked to study the issue. Given that many states in the West would resist implementing the kind of recommendations that NAS made for the Columbia, the states might be reluctant to make such a request.

To the extent that the NAS findings and recommendations are meaningful beyond Washington, might they indirectly affect water policy in other states? In other words, if other states regarded the Columbia River report as "good science" relevant to their own water resource issues, might they consider adopting its recommendations regarding water management in interstate river systems, water markets and conservation, requirements for new permits, and flexibility in water allocation and management? The most realistic answer seems to be no, given the western states' evident lack of interest in significant water law reform since at least the 1980s. On the other hand, there are at least three reasons to believe that some western states might be open and responsive to this type of scientific input in making decisions on water allocation and management.

First, there is now an important precedent: Washington asked NAS for a study of its water management options on the Columbia a remarkable step by a western state. Just as significantly, Ecology responded to the study with a proposal that was largely consistent with the NAS recommendations. It is certainly true that Washington is just one state, that its water policy is in some respects more progressive than that of other western states, and that its Columbia River management policy is still a work in progress which ultimately may not follow the recommendations. On the other hand, Washington is a fast-growing western state with a powerful agricultural community and a water code based on prior appropriation. If science can have a real water-policy impact in Washington, it potentially can do so elsewhere.

Second, the western states have repeatedly supported the use of science, and of adaptive management principles, in managing their natural resources. For example, the Western Governors' Association has espoused a set of principles for environmental management and decision making, one of which advocates the use of "a range of respected scientists and peer-reviewed science" to help establish basic facts and determine key areas of uncertainty. More directly relevant are the "Park City Principles," developed under the leadership of the Western Governors' Association and the Western States Water Council to guide future water resources decision making in the West. One of these principles states that water policies must be able to adapt to changing conditions, needs, and values; accommodate complexity; and allow managers to act in the face of uncertainty. This principle recognises the complex implications of the holistic approach to water management. Policies, and the managers who implement them, must provide flexibility to adjust as needed while ensuring predictability for those involved in and affected by water decisions. As decisions become more complicated, managers may want better and more certain information and guidelines. That will not always be possible, and the system must recognise and allow managers to act in the face of uncertainty.

While not mentioning science per se, this principle is generally consistent with the Report's endorsement of an adaptive management strategy for water resources that recognises the need to make decisions despite uncertainty and relies on science to test hypotheses and inform decision makers. Thus, at least in concept, "good science" is a key part of the western states' vision of good resource management. Third, science is just one of many factors that are nudging the western states toward changes in their standard approach to water allocation and management. The other factors include the increase in species listed under and protected by the ESA; the enormous population growth under way across the West; the ongoing economic transition, as agriculture and other traditional industries account for an ever-smaller slice of the region's economic pie; and the growing demand for environmental and recreational amenities, many of which rely on an adequate supply of water. None of these factors is primarily scientific in nature, but they all have exerted some pressure on the states to revise their water policies and practices, often (though not always) in ways that are consistent with the NAS recommendations. Thus, while science by itself is not powerful enough to bring about water policy reform at the state level, it may add some important influence to the nonscientific forces that create pressure for such reform in the West.

How would a state allocate water based on the NAS recommendations for good management? It would engage other states and other governmental entities, including tribes and federal agencies, in making decisions about new water withdrawals. It would look to existing water rights to provide the necessary supply for new uses through a combination of conservation projects and market-based transfers. It would consider potential impacts and risks to a range of public values, beyond the prospective economic benefits, in making decisions on new permits. And if new permits were issued, they would provide flexibility to make meaningful changes in the future as necessary to respond to changed conditions or new information.

Such proposals are certainly not new. Similar ideas have appeared in various water policy prescriptions over the past several years, such as the Park City Principles, the report of the Western Water Policy Review Advisory Commission, the report of the Longs Peak Working Group, and most recently the Interior Department's Water 2025. In general, such recommendations for water policy reform have had only minor practical effects on water policy in the western states. With the new Report, however, the recommendations have now been grounded in extensive scientific studies and delivered by an objective and highly credible scientific source. In a policy arena where conflicting values often result in gridlock, that scientific credibility should count for something. Of course, unlike the United States Supreme Court, NAS does not get the final word in matters of law. Nonetheless, "the Supreme Court of Science" speaks with a strong and authoritative voice, and its Columbia River Report may thus have a significant influence on water policy in Washington and beyond.

The Electricity-water Nexus and the Electric Utility Sector

During the drought of 2002, lawmakers in Idaho ruled that five large coal and gas-fired power plants should be denied water rights for cooling because they would deplete much needed freshwater for drinking and irrigation. In Nevada, the 1,580 megawatt (MW) coal-fired Mohave Generation Station was forced to close in 2005 due to lack of groundwater. A few years earlier, American National Power had to withdraw its application to build a 1,100 MW natural gas plant near Hillburn, New York, because it created a controversy concerning water rights. Far from being isolated examples, water issues have complicated power plant construction or operation in Arizona, Georgia, California, Colorado, Massachusetts, Missouri, New Mexico, North Carolina, Pennsylvania, Rhode Island, South Dakota, Tennessee, Texas, and Wisconsin.

Figure : Important connection in the energy-water nexus—agriculture.

The situation underscores a problem as pressing as it is invisible to many electric utilities, water planners, and even ordinary people:

burgeoning water use at conventional thermoelectric power plants. Water use for electric power plants increased five-fold from forty billion gallons per day in 1950 to 195 billion gallons per day in 2000. The average power plant in the United States uses about twenty-five gallons of water for every kilowatt hour (kWh) generated. Given that electric utilities produced 4,159,514 gigawatt hours (GWh) of power in 2007, these power plants ostensibly used 104 trillion gallons of water. This amount is enough to cover the entire country in two inches of water or to almost completely fill Lake Erie.

This chapter explores the consequence of the growing water needs of the U.S. electric utility industry, and suggests that lack of water during the summer months in many regions could complicate continued reliance on thermoelectric power plants that combust coal, oil, natural gas, and biomass (or utilise nuclear fission) to generate power. The importance of exploring the electricity-water nexus, its associated challenges, and its possible remedies is threefold.

First, a slew of government agencies and industry groups, including the National Research Council, U.S. Geologic Survey, U.S. Department of Energy, U.S. Department of Interior, Electric Power Research Institute, Sandia National Laboratory, National Energy Technology Laboratory (NETL), and the National Renewable Energy Laboratory, have recently issued reports focusing on the importance of water use at conventional power plants. These reports, however, mostly argue that better technologies will need to be developed in order to address the industry's growing water needs. The National Research Council calls on the federal government to increase research and development (R&D) funding for innovative energy technologies that utilise less water, while both Sandia and NETL discuss treating and reusing brackish water, capturing water vapour from power plants, and diffusion driven desalination as important technical options. Another study argues that the President should issue an Executive Order granting the Federal Energy Regulatory Commission (FERC) authority to designate select parts of the country "National Electricity-Water Crisis Areas." In December 2008, the Supreme Court heard arguments in Entergy Corporation v. Riverkeeper for and against strengthening the Environmental Protection Agency's regulations concerning the intake of water at conventional power plants, but the Court has not yet made its decision and the case concerns only the damage of power plant intake structures to fish and aquatic biodiversity. Senators

Bingaman and Murkowski even introduced legislation in early 2009 to commission a study on electricity-water problems and produce a roadmap, but their bill is uncertain to pass and again focuses on federal action. While focusing on federal research and national legislation is indeed important, equally significant is the role that electric utilities, public utility commissioners, and state regulators can undertake to avoid new thermoelectric power plant construction, invest in energy efficiency and renewable power resources, and alter electricity prices, either in conjunction with or independent of federal action.

Second, despite this collection of reports, existing electricity planners and water managers do not appear to be responding to electricity and water problems quickly or comprehensively. The Clean Air Task Force has also concluded that "water use and consumption have not been significant factors in decisions related to the permitting and siting of power plants." Peter Gleick has noted that "energy and water issues are rarely integrated into policy." Neither energy nor water planners are consequently trained to think about electricity and water in a systematic way. Electricity industry advocates continually downplay the importance of clean energy sources for minimising thermoelectric water consumption. When assessing solutions to reduce the water use from conventional generators, the Electric Power Research Institute, for example, mentions variable speed electric drives, advanced membranes, ozone disinfection, electroseparation, and freeze-thaw wastewater treatment as important water technologies, but not energy efficiency, demand side management, or renewables.

Conversely, an influential RAND report on water management focuses exclusively on tools such as "supply management" (including the location, development, and exploitation of new sources of water such as building new dams and control structures, desalination plants, arranging for inter-basin transfers of water, reclamation and reuse) and "demand management" (such as water quality matching, privatisation, and water pricing) but never once mentions energy policy tools or more efficient power plants. This chapter is an important call for more synergistic thinking that views electricity and water problems as interconnected, especially insofar that energy efficiency and renewable power plants can simultaneously reduce demand for electricity and improve the availability of water.

Third and finally, the challenges related to water scarcity and electricity are not confined to the United States. The Central Intelligence Agency believes that more than three billion people will be living in water-stressed regions around the world by 2015 (with a majority concentrated in North Africa and China). Water tables for major grain producing areas in northern China are dropping at a rate of five feet per year, and per capita water availability in India expected to drop fifty to seventy-five percent over the next decade. The American Museum of Natural History reports that about 900 million people spread across twenty-seven developing countries already lack adequate access to water. Thus, an exploration of how utilities in the United States may respond to electricity-water crises can offer policymakers insight into how the industry can address what is sure to become a pressing global dilemma.

The Electricity Water Nexus

Almost all conventional power plants, including coal, oil, natural gas, and nuclear facilities, employ one of three types of cooling cycles in their generation of electricity. Once-through cooling systems withdraw water from a source, circulate it, and return it to the surface body. Re-circulating or closed-loop systems withdraw water and then recycle it within the power system instead of discharging it. Dry cooling systems, which are not widely adopted, use air instead of water to cool power stations.

As their name implies, once-through cooling systems, or "open-loop" systems, only use water once as it passes through a condenser to absorb heat. After it passes through the plant, heated and treated water is then discharged downstream from its point of intake to a receiving body of water. Since such cooling systems release heated water back to the source, they can further contribute to evapourative loss by raising the temperature of receiving water bodies. Once-through systems are responsible for withdrawing ninety-one percent of the nation's water used for power plants, and are also utilised by more than half of the country's fleet of nuclear reactors.

Re-circulating or closed-loop systems, by recycling water, withdraw much less of it but tend to consume more. To maintain plant performance, water is frequently discharged from the system at regular intervals into a receiving body of water or collection pond, but is otherwise recycled as much as possible. Since it is being reused, the water requires more chemical treatment to eliminate naturally occurring salts and solids that accumulate as water evaporates. Closedloop systems also rely on greater amounts of water for cleaning and therefore return less water to the cooling cycle.

Figure : An emblem of the energy-water nexus: falling.

Dry-cooling, an approach that replaces evapourative cooling towers in closed-loop systems with cooling towers dependent entirely on air, works best in colder weather and in arid environments. Only a very small number of plants rely on dry cooling, since they lower plant efficiency and cost the most.

When taken together, the once-through, closed, and dry cooling systems in place at these power plants use a significant amount of water. (The term "use" encompasses both water consumption and water withdrawal). Using the most recently available data from the U.S. Geologic Survey, thermoelectric power plants used more than 195 billion of these gallons of water per day, or fortyseven percent of the nation's total, in 2000. According to the U.S. Geologic Survey, water withdrawals for thermoelectric generators differ greatly by state. When looked at geographically, Texas withdrew the largest amount of water; Illinois, Texas, and Tennessee accounted for twenty-two percent of all total freshwater withdrawals; and California and Florida accounted for more than forty percent of saline surface water withdrawals. This means that, on average, thermoelectric generators use more water than the entire country's agricultural and horticultural industry.

Such immense water needs produce equally immense concerns given the likelihood of future droughts and shortages, especially during the summer months. Even under normal conditions, water managers in thirty-six states anticipate shortages in the next ten years. Similarly, using a historical database of droughts going back to 1895, the U.S. Geologic Survey has predicted that almost one-fourth of the country will risk severe droughts by 2040. The most severely hit part of the country will be the West. As early as 2025, the U.S. Department of Interior cautions that "demands for water in many basins of the West exceed the available supply even in normal years." Given the intensity of the existing electricity industry's water needs, if it grows as predicted, it will directly trade off with the water available for drinking, industry, and agriculture. Utilising data from the U.S. Census Bureau, U.S. Energy Information Administration, U.S. Geologic Survey and National Oceanic and Atmospheric Administration, it appears that these tradeoffs will become most severe in twenty large metropolitan, areas. These regions of the country expect to add at least 2,700 MW of thermoelectric capacity and will experience population growth of at least 500 people per square mile. Thus, these regions will face water shortages of at least 1.52 inches in the summer by 2025.

Electricity Water Challenges in Eight Metropolitan Areas

To better illustrate many of the challenges facing electric utilities and water planners in these metropolitan areas, this section explores eight of them in detail: Atlanta, Charlotte, Chicago, Denver, Houston, Las Vegas, New York, and San Francisco. These eight regions plan to add a collective 129,828 MW of thermoelectric capacity between 2000 and 2025, power stations that would use 25.6 trillion gallons of water per year (or more than seventy billion gallons of water per day).

The dynamics of the water needs and consequences for each region will differ greatly. In some regions, such as Colorado, Georgia, and North Carolina, greater water use from power plants in 2025 could compete with the water needed for drinking, industrial manufacturing centres, and commercial enterprises. Water withdraws and consumption for power plants in Illinois and Nevada could deplete water from Lake Michigan and Lake Mead (respectively,) violating international law. In California and New York, greater power plant additions could compromise the stability of fisheries and accelerate the extinction of endangered species.

Figure : The greatest global water challenge according to new research.

In the Atlanta metropolitan area, Georgia Power, a subsidiary of Southern Company, intends to add 3,480 MW of thermoelectric capacity between 2000 and 2025. Georgia Power currently services 1.13 million customers in the Metro Atlanta region, yet their 16,000 MW portfolio is heavily water-intensive. About seventy-five percent of their fleet is coal powered, eighteen percent nuclear powered, six percent oil and gas powered, and one percent from hydroelectric sources. Indeed, more water will be lost as steam from Georgia Power's two nuclear plants than used by all residents of downtown Atlanta, Augusta, and Savannah combined. Within the state as a whole, thermoelectric plants use slightly more than half of all surface water, which then reduce drinking water supplies by reducing flows to Lake Lanier.

The most immediate consequence of increased thermoelectric water use in Atlanta will be tradeoffs with other major industrial and commercial water users in the region. These include Georgia-Pacific Corporation (one of the world's largest manufacturers of tissue, packaging, paper, pulp and building products), Mohawk Industries (the world's largest producer of flooring and carpets), and the city's water utility. The top commercial Atlanta customers for water included plants operated by die Coca-Cola Corporation, Pepsi Cola Corporation, Lockheed Martin Corporation, and Edwards Baking Corporation.

Pepsi's Gatorade plant, for instance, uses about five million gallons of water every month.

State policymakers seem to recognise the danger of water shortages, and a fierce legal battle has erupted. Georgia is fighting to hold back more water along its river basins and reservoirs, but Florida and Alabama argue that Georgia has mismanaged water resources and that extra Georgian withdrawals would dry up river flows that support out of state power plants, farms, fisheries, and industrial users. Alabama, for example, says that restrictions on water use in Georgia would impede electricity production at their Farley Nuclear Plant, also on die Chathoochee River, threatening power outages among 800,000 residents in three states. Tri-state water negotiations have so far only precipitated into eight active lawsuits, and Georgia's state assembly passed a resolution calling on the governor to set up a commission looking into having the border redrawn through die middle of Chattanooga, Tennessee. Resolutions were later introduced in both the state House and Senate to annex part of Tennessee to increase Georgia's access to water. The mayor of Chattanooga, who would lose half his city if Georgia's border was redrawn, sent a consignment of water bottles to Georgia lawmakers. He publicly announced it was better to "offer a cool, wet kiss of friendship radier than face a hot, angry legislator gone mad with thirst."

In Charlotte, Duke Energy Corporation reported plans to add 17,950 MW of diermoelectric capacity between 2000 and 2025. Duke Energy, the primary electric utility, relies on more than 28,000 MW of electricity capacity to meet the needs of 3.9 million citizens of Ohio, Kentucky, Indiana, North Carolina, and South Carolina (die Carolinas account for 2.3 million of its customers). Of its total capacity, thirty-nine percent is coal-fired, thirty-seven percent is nuclear, thirteen percent is hydroelectric, and eleven percent oil and natural gas meaning that their entire portfolio is water-intensive.. Duke already announced in March 2008 that it needed to import 520 MW of power outside of the region to ensure continuation of service during a prolonged drought. One town in North Carolina was so dry mat water had to be imported by fire truck. In 2008, die Summer nuclear plant (near Columbia) was at such a "critical point" that operators openly discussed having to shut it down for lack of water. The Harris nuclear reactor near Raleigh obtains water from Harris Lake, which was a scant 3.5 feet above the limit that the plant could operate. Most of Duke's power plants draw their cooling water from the Santee River Watershed, which includes the Catawba- Wateree River, a water source that has earned the title "America's Most Endangered River for 2008. This "Most Endangered River" basin, however, is about the only place to situate new thermoelectric power plants, and the associated water use with these capacity additions could exacerbate drought (at best) and risk interstate litigation and agricultural collapse (at worst). Low water levels along the river leave bottom dwelling organisms such as clams and mussels stranded, and tend to induce algal blooms that contaminate water supplies. Low water levels also degrade drinking water infrastructure, since they diminish revenues for water utilities (meaning they have less earnings available for maintenance,) and dry and crack soil leading to pipeline malfunctions and spills.

In Chicago, electric utility planners intend to add 40,386 MW between 2000 and 2025. Chicago is formally served by Commercial Edison, a subsidiary of Exelon Corporation. Exelon, one of the largest electric utility providers in the nation, serves 5.2 customers, but under a $4.8 billion deal sold many of its power plants to Midwest Generation (which generates electricity but only sells it on the wholesale market). Commercial Edison, now the nation's largest supplier of nuclear power (and the world's third largest supplier), operates one of the most water-intensive fleets in the electricity industry. The total amount of thermoelectric power generation in the area is about 12,649 MW, drawing mostly from a mix of local rivers, Lake Michigan, and Powerton Lake.

The most immediate impact to future thermoelectric withdrawals from Lake Michigan could be a violation of domestic and international law. The Supreme Court decided in 1996 that the State of Illinois had to limit its diversion and use of water from Lake Michigan. Under the ruling, Illinois must legally reduce its water usage over the next fourteen years - not increase it for power plants. Furthermore, since the Great Lakes contain twenty percent of the world's freshwater supply and ninety-five percent of the freshwater for the United States, they are strenuously protected under international law. The 1909 International Boundary Waters Treaty, signed between Canada and the United States, specifically governs water resource use on Lake Michigan. The most recent agreements, signed by all Great Lakes State Governors and Canada's Provincial Premiers in December 2005, recognise that "the Waters of the Basin are a shared public treasure and the States and Provinces as stewards have a shared duty to protect, conserve and manage these renewable but finite Waters." Thus, deploying more thermoelectric power plants could defy domestic court rulings and international treaties.

In the Denver metropolitan area, Xcel Energy may build 4,503 MW of thermoelectric capacity between 2000 and 2025. 'Xcel operates three power generating facilities in Denver - the Cherokee, Zuni, and Arapahoe Plants - all which use water from the South Platte River. Located south of downtown, the Arapahoe Station is a 156 MW coal-fired plant that uses about 586 million gallons of water annually from the South Platte River. The Zuni Station, closer to downtown, is a 107 MW natural gas plant that uses 98 million gallons of water annually from the South Platte River. The Cherokee Station, north of downtown, is a 717 MW coal-fired plant that uses 2.4 billion gallons of water from the South Platte River annually (along with about 5,200 acre-feet of recycled water from Denver Water's Reuse Water Plant).

Because the South Platte River is the primary water source for the region, increased power plant withdrawals and consumption could seriously deplete the water needed for households and businesses (about forty-seven percent of Denver's water is used by single family homes). Water use from any additional power plants would have to be balanced against the needs of the many downstream users along the South Platte. A severe shortage could risk economic and environmental impoverishment, since the municipal drinking water system in Eastern Colorado and farming and ranching activities in Nebraska rely on the river.

In Houston, electricity demand could grow to need 26,989 MW of new thermoelectric capacity by 2025. The situation with retail electricity access in the Houston area is more complicated than many other cities (although the construction and dispatch of generation remain roughly the same). In June 1999, Texas Senate Bill SB 7 made retail electric competition state law and phased-in "retail choice" of electricity on January 1, 2002." The legislation forced incumbent Texan utilities such as TXU and Reliant Energy to restructure then assets into separate retail marketing, generation, transmission, and distribution companies. The bill attempted to lower prices and limited the market power of existing suppliers by unbundling transmission, distribution, and generation; it also prohibited any single utility from owning more than fifteen percent of generation within a service region. Thus, the electricity market serving Houston is complex. More than 200 different electricity plans are available to Houston residents and service is offered from sixty-eight competitive retailers, including eight large utilities Champion Energy, CPL Retail Energy, Direct Energy, Gexa Energy, Green Mountain Energy, Reliant Energy, Spark Energy, and TXU Energy that generate and distribute power from five distinct zones all over the state. Taken together, these power providers operate about 95,400 MW of capacity, forty-eight percent fuelled by natural gas, thirty-nine percent by coal, eleven percent from nuclear, and three percent from "other" sources. The largest of these is the "wires only" utility Centrepoint Energy, which maintains the T&D infrastructure serving 5,000 square miles around the city (although it does not generate a single kwh of power). Centre Point serves two million customers in Houston and operated 3,600 miles of transmission lines, 226 substations, twelve service centres, and 43,000 miles of distribution lines to deliver 75.9 billion kwh in 2006.

Increasing the number of power plants in the Houston vicinity, however, could reduce the water needed for drinking and agriculture. In the past, Houston depended primarily on groundwater to meet eighty percent of its supply, but rapid depletion has lowered that amount to only sixty-seven percent today, forcing the city to take more water from the Trinity, San Leon, and San Jacinto Rivers (along with the reservoirs they support). With Houston water planners predicting rising demands for drinking water, there may not be enough water for new power plants. Surface water upstream from Houston is also needed to irrigate agriculture. During the last serious water shortage caused by a prolonged drought in 1996, the agricultural sector was the first to suffer as water was diverted to supply power plants and drinking water systems.

Las Vegas, Nevada

Population is growing so quickly in Nevada that more than 20,000 MW of thermoelectric capacity may be built in Las Vegas between 2000 and 2025. Because of the rapid growth in population, Nevada Power Company, the primary electric utility, is also one of the fastest growing utilities in the country. Nevada Power's service area covers 807,000 customers spread across 4,500 square miles in southern Nevada, including the cities of Las Vegas, North Las Vegas, and Henderson. It operates 35,990 miles of T&D lines and has invested

$5.4 billion in about 5,623 MW of mostly natural gas and coal-fired power plants, which it uses to meet 54.3 percent of the region's power needs. The shortfall is made up from purchased power imported from Arizona, Utah, and other parts of the state.

The water needs for this portfolio of power technologies is significant. Nevada Power's fleet of plants currently use water from a combination of local reservoirs, underground aquifers and wells, and the Colorado River, with a significant amount coming from the beleaguered Lake Mead. According to water use profiles reported by the Nevada Division of Environmental Protection, the 465 MW El Dorado gas power plant uses 176,100 gallons of water per day from Lake Mead. The almost completed Copper Mountain Power Plant, a 660 MW natural gas plant soon to be operating in Clark County, will use 75,000 gallons of potable water per day from Lake Mead. The Reid Gardner Station, a four-unit 650 MW coal fired power plant in the Moapa Valley, uses groundwater that would otherwise replenish Lake Mead. The power plant consumes about 286,000 gallons of water per day from the Muddy River (which flows into Lake Mead), but has also contaminated local aquifers and wells. The Reid Gardner station's cooling ponds and waste landfills have leaked sulfates and toxic pollutants into the watershed. Groundwater underlying the power plant has become so contaminated that it exceeds federal safety standards for no less than twelve criteria pollutants including arsenic, chromium, lead, and selenium.

If more thermoelectric power plants are built as planned, the most immediate impact would be greater water shortages for the Las Vegas Valley Water District and further reductions in Lake Mead, which is already experiencing and "extreme hydrological drought."" Researchers from the Federal Bureau of Land Reclamation estimate that the Lake Mead water system is losing 326 billion gallons of water per year, or enough to supply eight million people, a loss so significant it can be seen by satellite images from space. If new power plants further tap Lake Mead, the result could be a full blown agricultural disaster. Lake Mead directly irrigates about a million acres of farmland in southern California's Imperial Valley, "and another half million acres in northern Mexico as part of [an] international water treaty."

New York, New York

Planned thermoelectric capacity additions for New York could surpass 11,600 MW by 2025. The major electric utility directly serving

New York City is Consolidated Edison, which provides electric service to most of the City (excluding a small part of Queens) and Westchester County. The utility operates about 1,739 MW of its own capacity (mostly oil and gas plants) to supply 10.5 percent of its power, but imports the rest from competitive suppliers through an integrated bulk power system known as the New York Control Area. About half electricity generated within the city comes from power plants in Queens County, and a majority of this power comes from the 2,450 MW Ravenswood natural gas power plant on the East River. Consolidated Edison takes its generated and purchased power and delivers it to about three million customers through 94,000 miles of underground electric cables - the largest underground system in the world and enough cable to wrap around the Earth more than three times.

The water needs for the portfolio of power generators that must supply power within Consolidated Edison and the New York Control Area are substantial, but they do not tradeoff directly with the City's water supply. The City's water comes from nineteen reservoirs, three controlled lakes, and about 300 miles of aqueducts spanning the Catskill Mountains to Westchester County. Instead, new power plant additions would likely consume and withdraw water from the Hudson River, where seven facilities constituting 6,691 MW of capacity - Bethlehem Energy Centre (previously the Albany Steam Station), Danskammer Generating Station, Rosetón Power Plant, Indian Point Energy Centre, Lovett Power Plant, Bowline Power Plant, and the LRT Power Plant on 59th Street - already use 6.1 billion gallons of water directly from the Hudson for coolant every day.

New power plants could consequently have a devastating impact on local fisheries and ecosystems through the discharge of heated effluent, entrainment, and impingement. A series of extensive fishery surveys along the Hudson, including the Long River Survey (assessing egg and larval densities), Fall Shoals Survey (assessing densities of juvenile fish populations), and the Beach Seine Survey (assessing the abundance of fish communities) determined that thermoelectric power plants were devastating freshwater fisheries in the early 1970s. The federal government passed extensive regulations to limit the damage, but utility restructuring this past decade has renewed concern that electric utilities, more focused on competition, will be less focused on environmental compliance. Indeed, one 2006 study has already warned that additional thermoelectric power additions along the Hudson could increase mortality of Striped Bass (already weakened from current power plants), bay anchovy (already suffering fifty percent year class reductions due to power plants), and Atlantic Tomcod (at stress due to any further increases in water temperature). A similar 2007 study warned that the withdrawal of cooling water for new thermoelectric plants would have:

"profound impacts on aquatic environments" [including] "reductions of phytoplankton, zooplankton, fish, and shellfish, including the loss of threatened and endangered species; damage to aquatic organisms, including important elements of the food chain; reduction of populations and their compensatory reserves; losses of commercial and recreational fisheries; and stresses to overall communities and ecosystems."

San Francisco, California

In San Francisco, electricity demand could grow to justify the construction of 4,759 MW of thermoelectric capacity between 2000 and 2025. Both competitive power producers and electric utilities operate in the San Francisco Bay Area. The Mirant Corporation, an independent power producer, operates the 674 MW Contra Costa natural gas plant, the 1,311 MW Pittsburg natural gas plant (both situated near the confluence of the Sacramento and San Joaquin Rivers), and the 362 MW Potrero natural gas plant (which uses cooling water directly from the San Francisco Bay). Pacific Gas and Electric Company (PG&E) also operates the 530 MW natural gas-fired Gateway Generating Station near Antioch. The plant receives its water from the Delta at the confluence of the Sacramento and San Joaquin Rivers. PG&E's 2,200 MW Diablo Canyon Nuclear Power Plant, located in Avila Beach, uses water from the Pacific Ocean.

Given the region's unique hydrology, the immediate effect of additional water use from new thermoelectric power plants could be drought, shortages of drinking water, and species extinction. Most of power plants in the Bay Area draw water from the Sacramento-San Joaquin Delta, a massive, natural, inverted river delta, where numerous waterways converge downstream, eventually flowing into the Suisun Bay and the Upper San Francisco Bay (where they proceed to flow into the Pacific Ocean under the Golden Gate Bridge). The same water system replenishing the rest of California, therefore, is the one that power plants predominately use. The water system is instrumental in distributing water to twenty-five million Californians extending beyond San Francisco through the Central Coast to Los Angeles and San Diego. About 2.5 million acres of productive farmland rely on the Sacramento-San Joaquin Delta for irrigation, and the California Draught Preparedness Organisation has already warned that water supplies are running dangerously low. Further contributing to a shortage of water, a 2007 District Court ruling protecting the endangered Delta Smelt effectively cut California's drinking and water irrigation supplies by thirty-five percent. Almost one month after the court decision, the Coalition for a Sustainable Delta filed a lawsuit against the Mirant Corporation, alleging that their Contra Costa and Pittsburg natural gas power plants harmed the Delta Smelt by using more than a billion gallons of water a day to cool steam turbines. In April 2008, further restrictions on water withdrawals were implemented out of concerns that rapid declines in Delta water were harming the Sacramento River Chinook Salmon. New conventional power plant additions, it appears, could risk possible species extinction in addition to drought and water shortages.

Electric Utilities and Regulators

What can regulators in these cities and states, along with the eight utilities most at risk and the dozens of others possibly like them, do in the face of these electricity-water challenges? While many disputes over water in the West and Southeast are interstate, local regulators and electric utilities are still exceptionally positioned to respond to such risks within their jurisdiction. City and state regulators are often much more attuned to the needs and dynamics of their constituents than federal policymakers, and can respond more quickly and efficiently to local problems. Electric utilities serve as direct intermediaries between electricity consumers and power providers and generators; have established supply,metreing and billing relationships with their customers; are more flexible than government agencies and can quickly tailor programs; and possess the most knowledge about electricity production and use in their area.

This section argues that while a cornucopia of different technologies and mechanisms are available to regulators and electricity utilities, a combination of six could be very effective at avoiding future water shortages: increasing research and development funding for alternative power plant cooling cycles, placing a moratorium on thermoelectric power generation, strongly promoting energy efficiency and demand-side management, rapidly deploying wind turbines and solar photovoltaic panels, changing electricity prices and giving electricity customers more feedback and information, and asking the federal government (or forming interstate organisations) to designate select regions of the country "Electricity-Water Crisis Areas."

Figure : Regulators and buswork inside an electric utility substation.

Increase Funding for R&D on Alternative Cooling Cycles

The U.S. Department of Energy, through separate programs at the National Energy Technology Laboratory and Sandia National Laboratory, has begun to look into making conventional power plants more water-efficient, and a number of emerging technologies can greatly reduce water use. Researchers working at these programs, for example, have investigated treating and reusing "impaired," "nonpotable," "produced," "brackish," "reclaimed," or "gray" water to cool power plants. The most common applications include using secondary treated municipal waste water, passively treated coal mine drainage, and ash pond effluent. Fifty-seven power plants, mostly in the arid Western part of the United States, already rely on cooling cycles that utilise reclaimed water, and while abundant sources of reclaimed water are available in Alaska, California, Kansas, Louisiana,

Oklahoma, Texas, and Wyoming (accounting for 90.1 percent of produced water), it would be a prohibitively costly for forty-three states. Of the produced water in these seven states, only thirty-seven percent is of the quality needed to run in power plants.

Another option being explored is enabling power plants to produce some of their own water, either through capturing water vapour from flu gas or using the thermal discharges from power plants to desalinate water. Water is naturally present in all deposits of coal, constituting as much as sixty percent of its weight. The coal combustion process thus releases water vapour which can be recovered from flu gas using liquid desiccant-based absorption systems or modified electrostatic precipitators. Engineers at NETL expect that such capture technologies could reduce five percent of evapourative water loss at power plants. Diffusion driven desalination, a process that uses the excess waste heat from power plants to produce distilled water, can also minimise the water needs of power plants situated in coastal areas.

Thermoelectric power plants, in other words, need not always consume and withdraw water as they do today, and opportunities exist to substantially improve the efficiency of thermoelectric cooling cycles. While none of the technologies discussed here are yet available at a commercial scale, doubling or tripling the funding for the research programs at Sandia and NETL could provide the technological breakthroughs needed to greatly reduce water use at power plants in the future.

Place a Moratorium on New Thermoelectric Power Generation

Perhaps the simplest response electric utilities can take is to stop building new thermoelectric generation in areas where water shortages are expected to occur, or water prices anticipated to rise rapidly. As the two next sections suggest, existing thermoelectric and hydroelectric capacity would be sufficient to meet existing power needs but energy efficiency, wind, and solar could offset any expected increases in electricity demand.

Using these three tools to offset thermoelectric capacity additions is important because drought and flood are a normal, recurring part of the North American hydrologie cycle. Even though meteorological droughts, identified by a lack of measured precipitation, are difficult to predict and can last months to decades, every part of the country has experienced severe or extreme drought conditions at least once since 1896 - with about half of the country suffering drought conditions ten to fifteen percent of the time. Given the likelihood of future water shortages, utilities could justify transitioning away from water intensive thermoelectric generation.

The addition of new conventional power plants therefore has two inherent water-related risks that suggest electric utilities should no longer construct them: they are unable to withdrawal water needed for normal operation in times of scarcity, and can cause and worsen existing water shortages when their fuel cycles consume water. Thermoelectric power plants running on coal, natural gas, oil, and uranium require immense amounts of water to cool the combustion process. They withdraw trillions of gallons of water from our rivers and streams. They consume billions of gallons of water from local aquifers and lakes. A moratorium would therefore offer a very effective tool for preserving water resources.

One possible objection to a moratorium would be that future increases in electricity demand can only be reliably met by fossil-fuelled and nuclear baseload power plants. While this concern is a legitimate one, the next two sections show that the promotion of energy efficiency, demand side management, and improved feedback to electricity customers could offset the need to build any new thermoelectric capacity.

Furthermore, attaching wind turbines to pumped hydro and compressed air energy storage systems can improve their capacity factor to above seventy percent, making them "functionally equivalent to a conventional baseload plant." Paul Denholm found that generating base-load power from a hybrid wind/compressed air energy storage system would add only about 0.70/kWh to its cost of producing electricity, and that converting natural gas plants to biomass generators to backup wind farms would add only about 0.2 0/kWh, two relatively inexpensive options to create completely renewable base-load units. Contrary to some proclamations stating otherwise, "a baseload wind system," notes Denholm, "can produce a stable, reliable output that can replace a conventional fossil or nuclear baseload plant."

Similarly, Vasilis Fthenakis and his colleagues from Brookhaven National Laboratory, the Renewable Energy Research Institute, and the Institute for Analysis of Solar Energy recently concluded that with new advances in compressed air energy storage make it "clearly feasible to replace the present fossil fuel energy infrastructure in the

[United States] with renewables." Researchers at the Georgia Institute of Technology and the Virginia Polytechnic Institute & State University even found that when coupled with a rigorous energy efficiency and demand management program, solar panels could completely displace the electricity currently coming from the two GW Indian Point nuclear facilities in New York at a cost less than building a new nuclear power plant.

Promote Energy Efficiency and Demand-Side Management

To offset the risks of placing a moratorium on future thermoelectric generators, electric utilities should rigorously implement energy efficiency and demand-side management programs (DSM). Curiously, such action would not only help address impending electricity related water shortages, but would also improve energy security, lower electricity and water prices, and enhance reliability. For the historical record suggests that energy efficiency, DSM, and load management practices represent the most feasible way of responding to increases in electricity demand. Increasing energy efficiency, one study concluded, "is generally the largest, least expensive, most benign, most quickly déployable, least visible, least understood, and most neglected way to provide energy services." Or, as Jon Wellinghoff, the Commissioner of the FERC, put it, "the potential benefits from the incorporation of demand response into wholesale markets indicate that a considerable margin of gain is possible from accelerating such activity."

The Department of Energy (DOE) recently calculated the benefits of DSM and found that it lowers wholesale electricity prices as costly power plants are displaced and total demand on the system decreases. Generating peak electricity is extremely expensive, often exceeding $5,000 to $10,000 per installed kW (meaning a 100 MW plant can cost $750 million to build and require seventyfive million dollars per year to operate), implying that DSM should be profitable for all utilities. Eric Hirst estimated the costs and benefits of DSM programs for three types of utilities: a "surplus" utility, an "average" utility, and a "deficit" utility.

Surplus utilities are those with excess capacity and few planned retirements, as well as slow projected growth in fossil fuel prices and incomes". Deficit utilities have little excess capacity, many planned retirements, and rapid growth in prices and incomes. Average utilities fall in the middle. Hirst found in each case that DSM programs raise electricity prices but reduce electricity costs, and the overall percentage reduction in cost far exceeds the increase in price. He found 2-1 cost benefits for the surplus utility (i.e., every dollar spent on DSM yielded two dollars of savings), 5-1 cost benefits for the average utility, and 8-1 cost benefits for the deficit utility. In other words, energy efficiency may cause electricity prices (per kWh) to go up slightly, but because people and companies use fewer kWhs, their bills actually go down. For some utilities, cost savings can be as high as eight dollars for every one dollar invested in DSM.

For these reasons, a few states have experimented with large-scale DSM programs. In the state of New York, cost-effective DSM and load management policies have saved more than 1,000 gigawatt hours (GWh) of electricity and displaced 880 MW of peak demand in the past five years. Montana, Idaho, and Oregon conserved energy totalling more than 50,000 GWh per year in the mid-1990s with a retail value of two billion dollars to consumers. California's DSM programs operating between 1990 and 1992 delivered 112 percent of planned energy savings, meaning they cost-effectively saved more than electricity than intended. The Massachusetts Department of Energy Resources, discovered their load management program lowered participants' electricity costs by twenty million dollars in 1999, with benefits to Massachusetts customers exceeding six million dollars in just thirteen hours on one high-cost day.

Notwithstanding these impressive gains, there is much more potential in energy efficiency and DSM than some ever imagined. The National Association of Regulatory Utility Commissioners (NARUC) found cost-effective energy efficiency potential in all regions of the country, with the most untapped potential in the Northeast and South, where electricity costs are highest (meaning energy efficiency efforts are more economical than areas where energy is cheaper). Another study projected that cost-effective energy efficiency programs could reduce consumption by around one trillion kWh by 2020, offsetting almost all projected growth in electricity use - and the needed capacity additions to achieve it. The Alliance to Save Energy found that aggressive investments in energy efficiency could free up enough electricity to eliminate the need to construct more than 1,300 power plants in the next twenty years. One study projected that a national DSM program aimed at reducing peak demand by just five percent would yield three billion dollars in net generation, transmission, and distribution savings per year and displace some 625 infrequently used peaking plants and associated delivery infrastructure. Other broader studies confirm DSM's cost-effectiveness. The International Energy Agency reviewed forty large-scale commercial DSM programs found that they saved electricity at an average cost of 2.1 to 3.0 eVkWh. Similarly, the Institute of Electrical and Electronics Engineers found an average cost of 2.6 0/kWh for demand-side management, load management, and energy efficiency programs in Vermont166. When thinking about these numbers, readers are asked to consider that one kWh saved is more valuable than one kWh generated. One thousand MW of energy efficiency and 1,000 MW of large-scale generation are not equivalent. Richard Cowart found that every dollar invested in energy efficiency:

- Mitigated against uncertainty and lowered load, wear, and maintenance needs on the entire fossil fuel chain, even in hours when reliability problems were not anticipated by system managers;
- Depressed the costs of locally used fuels such as oil, coal, and natural gas;
- Reduced demand across peak hours, the most expensive times to produce power;
- Lessened costly pollutants and emissions from generators;
- Improved the reliability of existing generators;
- Moderated transmission congestion problems;
- Operated automatically through customers coincident with the use of underlying equipment or load, meaning they are always "on" without delay or the needed intervention by system operators to schedule or purchase the resource.

Discounting all of these benefits and looking purely at reductions in peak demand, the New York ISO determines a reserve criterion of eighteen percent during times of peak demand to ensure overall system reliability. Accordingly, each 1.0 MWh of peak demand that customers avoid through energy efficiency means that utilities can subtract 1.18 MWh of total capacity needed. Quite literally, every single kWh of peak electricity avoided through energy efficiency equates to 1.18 kWh of avoided supply. While this calculation would not necessarily hold true for the displacement of base-load generators, since most peaking power plants rely on natural gas the water savings could still be substantial. Natural gas units are the third most waterintensive of all power plants, using about fourteen gallons of water per kWh (second to coal at thirty-six gallons per kWh and nuclear power at forty-three gallons per kWh).

Furthermore, a majority of energy efficiency savings can be accomplished with small and targeted programs. In the ISO-New England service area, for example, about nine percent of the system's total generating capacity is tapped one percent of the time. The price of power during this most expensive one percent accounts for sixteen percent of the total annual dollars spent on the spot market. These numbers reveal that a very small fraction of supply accounts for a large fraction of total cost. Nationwide, Cowart calculated that just 0.4 percent of industrial customers account for thirty percent of total demand. Consequently, relatively small DSM programs directed at a miniscule proportion of the nation's electricity customers can produce mammoth benefits in terms of total demand reductions.

Finally, a significant amount of energy efficiency and DSM potential exists in the eight metropolitan areas identified above. Georgia Power could cut projected load by thirty-three percent for residential and commercial customers (and twenty-six percent for industrial customers) by 2018. The same study found an achievable potential for energy efficiency programs - that is, energy that could be saved if existing programs were funded to their full potential - of 2,509 GWh per year. Another assessment conducted by the Governor's office estimated that twenty-three to twenty-four percent of cumulative load growth could be displaced by cost-effective energy efficiency programs (amounting to the equivalent of about 1,771 MW of displaced capacity).

In North Carolina, Duke Energy has acknowledged that their "save-a-watt" program could displace 1,318 MW of peak capacity by 2012 if properly funded. Similarly, Lawrence Berkeley National Laboratory researchers estimated that Chicago had at least 3,456 MW of untapped DSM potential. Environment Illinois, a nonprofit group, concluded that home weatherisation programs, enforcement of energy star appliance standards, and light bulb replacements could cost-effectively save the state more than 2,100 GWh, displace the need for fourteen billion cubic feet of natural gas, and cut home energy use by twenty percent. If coupled with more aggressive retrofitting schemes and other investments, household energy consumption would be reduced by forty percent (and with technological improvements, ninety percent). In Colorado, Xcel Energy admitted that it could ramp up its DSM program to provide an additional 200 GWh (equivalent to eighty MW of peak potential) if properly incentivised.

An independent report from KEMA found that the state of Colorado at large had at least 2,031 MW of technical energy efficiency potential. Texas has 40,000 MW of energy efficiency potential characterised by demand response, solar PV, and improved building codes, appliance standards, and public energy efficiency initiatives. These alternative sources could meet 107 percent of the projected growth in summer peak demand by 2013 and displace thirty-three percent of peak load by 2023.

A DSM program in Texas would not only displace conventional generation, but would also result in utility bill savings of seventy-three billion dollars at an avoided cost of 4.50/kWh, much cheaper than the state-average retail electric price of 9.IeTkWh. And, in Nevada, fully funding just nine energy efficiency programs could displace the need to construct 5,571 MW of capacity in the Sierra Pacific and Nevada Power service areas by 2020. In New York, the National Research Council, along with a study led by researchers at the Georgia Institute of Technology, found a significant amount of programmable energy efficiency potential within New York City. That is, existing programs would only need to be properly funded to achieve roughly 4,461 GWh of potential by 2012.

Both studies show that residential energy efficiency improvements in lighting, cooling, refrigeration, electronics, space heating, and hot water heating, along with commercial improvements in lighting, refrigeration, cooling, ventilation, office equipment, water heating, space heating, and building controls, could cost-effectively displace the need to build 550 MW of peak capacity by 2015. The studies estimated that an additional 300 MW of peak capacity could be displaced by aggressive demand response programs. The Natural Resources Defence Council and Pace University Law School Energy Centre (now the Pace Energy & Climate Centre) estimated even greater potential in the New York City area, and concluded that savings of up to 3,032 MW peak demand could be achieved by aggressive energy efficiency programs within two years. Finally, the San Francisco Bay Area Economic Forum calculated that scaling up load management programs in the region could displace 2,500 additional MW of thermoelectric capacity.

Deploy Wind Farms and Solar Photovoltaic (PV) Panels

In situations where energy efficiency and DSM programs are unable to completely offset the need to construct new thermoelectric power plants, utilities could rely on wind turbines and solar panels to produce electricity.

These two technologies use almost no water to generate electricity, and need only a very small amount for cleaning and maintenance. Even more remarkably, looking at the marginal levelised cost of building power plants in 2009 - that is, the cost of constructing, operating, maintaining, and fuelling a new facility - offshore and onshore wind turbines produce electricity for between 2.6 and 5.6 0/kWh, making them two of the six cheapest sources of power. Solar PV is the most expensive at 39 0/kWh, but not far behind expensive peaking plants that cost between 32.5 and 35.6 0/kWh to operate. Wind, in other words, is already cheap, and solar (which is getting cheaper) is nearing parity with natural gas peaking facilities.

Nationally, commercially available wind and solar photovoltaic power generators could provide almost 3,000 GW of capacity - roughly three times the country's existing capacity - by 2010 if only they were built. (Building these wind and solar power facilities would also have the additional benefit of providing high-paying manufacturing jobs in those regions of the country hardest hit by the financial crisis). Again, while the following estimates and projections should be taken in context, the eight metropolitan regions discussed in Part II have substantial amounts of wind and solar potential.

In Georgia, solar PV systems could provide at least 104 MW of cost effective peaking power by 2010. In Atlanta, 13,223 MW of peak capacity could be displaced by placing solar arrays on just commercial and residential buildings. Moreover, certified wind maps project more than 10,000 MW of offshore wind power could be imported to Atlanta. In North Carolina, an independent consulting team found 9,600 MW of onshore wind potential, with a "practical" amount of 1,500 MW, that could be installed in Duke Energy's service territory to produce 3,900 GWh of electricity pear year. Indeed, researchers at Stanford University and the University of Delaware projected that offshore wind energy near the Mid-Atlantic coast could supply 330,000 MW of power - enough to meet the energy needs of nine states from Massachusetts to North Carolina (including D.C.) with electricity left over to support a fifty percent increase in demand.

In Chicago, the DOE confirmed that Illinois and the region surrounding the city had at least 9,000 MW of potential onshore wind power at only five of the state's prime sites. These areas represent about 1.2 percent of state land and each square kilometre of windy land could support about five MW of installed wind capacity. Offshore wind potential in Lake Michigan is even greater. A report partially funded by the Wisconsin Focus on Energy Program, after sampling hundreds of wind sites and assessing lake depth, concluded that the mid-lake plateau would be "excellent" for offshore wind farms with steady and strong winds, and that the Lake has energy potential "in excess of 10,000 MW." The estimate was confirmed by an independent and interdisciplinary analysis of potential offshore sites which found that 9,700 MW could be developed.

Colorado's combination of high mountains and flat plains creates a bounty of wind resources, with strong and consistent winds coursing through the western part of the state and Denver. The Governor's Energy Office has noted that a significant amount of this wind - about 93,000 MW - could be converted into electricity. The DOE found more potential, and noted that the state has about 170,000 MW of potential wind capacity (with the capability to produce 288 billion kWh of electricity per year), even when making exclusions for land use and environmental permits. The Hewlett Foundation also projected that eighty-three million MWh per year of solar PV resources could be deployed within the state, or about twice as much electricity as consumed state-wide.

Three of the five largest wind farms in the nation are located in Texas: the state holds the record for the world's largest wind farm (the 736 MW Horse Hollow Wind Energy Centre), and it is the site for the nation's second-largest wind farm (the 505 MW Sweetwater Wind Project). The potential, nonetheless, has hardly been exhausted, and the Texas Public Utility Commission estimates that as much as 25,000 MW of wind power could flow to urban centres by 2012. The Union of Concerned Scientists projects that die state could add another 17,800 MW of wind by 2025.

In Nevada, after assessing land use patterns and favourable wind sites, researchers discovered a large concentration of achievable wind energy near Las Vegas. The University of Nevada calculated that state had the potential for more than fifty million MWh of wind power, or the equivalent of 3,240 MW of installed capacity in the three counties surrounding Las Vegas. Solar PV resources for Southern Nevada are excellent: about 7,000 to 7,500 watts hours per square metre, making Las Vegas "one of the best sources for this type of generation in the world." Researchers at the Hewlett Foundation and the Energy Foundation projected that the tri-county area surrounding Las Vegas could provide 11,700 MW of power from solar PV systems. New York City has an equally impressive amount of building integrated solar PV potential. Researchers at the Centre for Sustainable Energy at Bronx Community College projected that solar PV systems mounted on rooftops and building façades could supply eighteen percent of the city's electricity by 2022. Each square foot of New York City receives the equivalent of 160 kWh of sunlight per year, enough commercial and residential roof space to host between 8,500 MW and 15,700 MW of PV installations. Researchers at the University of Albany estimated that 2,431 MW of solar PV potential exists for the state of New York on large parking lots alone.

Lastly, the San Francisco has estimated that it has at least thirty MW of small-scale wind generation in the Bay Area. The nonpartisan California Energy Commission (CEC) concluded that the Bay Area could rapidly deploy 4,000 MW of wind potential in the form of large onshore wind farms in only a few years. Stanford University researchers projected that between twenty-six percent and 12 percent of California's entire electricity needs could be provided with offshore wind. A single 2,385 MW wind farm sited on the Northern California coast, for instance, could produce about 9.7 TWh of electricity, enough to displace 5.6 percent of the state's fossil fuelled generation (the authors comment that the 9.7 TWh estimate is on the low end of the potential for the Bay Area, with a high end of twenty-seven TWh).

Researchers at the Golden Gate University Law School estimated annual PV output for San Francisco at 550 to 1300 GWh per year. At the high end, this is comparable to current residential (and industrial) electricity use in San Francisco. At the low end, the researchers estimated that PV output could displace 600 MW of peaking power plants. The researchers noted that over the thirty year period that they analysed, the standard deviation of annual average solar radiation was less than 2.5 percent of its average value, suggesting that solar resources would be very reliable. They also noted that these PV systems could be integrated into buildings and rooftops, meaning they would have no new requirements for land. The CEC analysed the state's solar PV resources and found that since they could be deployed virtually anywhere in the state, their technical potential exceeded seventeen million MW of capacity. If applied only to existing residential and commercial rooftops, the CEC concluded solar PV potential exceeded 74,000 MW of capacity.

Change Electricity Prices and Improve Information

A moratorium on thermoelectric generation, energy efficiency and DSM programs, and wind and solar PV deployment should be supplemented with utility efforts to alter electricity prices and provide feedback and information to electricity customers. Three changes would be most significant: more accurate electricity pricing, altered electricity billing practices, and a utility-wide information program to educate consumers.

Consumers are generally unaware of daily, weekly, and seasonal changes in electricity price, and instead see only a monthly electricity bill. They thus consume electricity indiscriminately. Utility programs that reflect time-of-use through "real-time," "interval metreing," "time-of-use", or "seasonal" rates could show customers how electricity production and consumption varies according to the time of day, week, and month. Most electricity bills combine charges into a lump sum, making it difficult for consumers to tell how much of the bill results from the individual use of appliances or technologies, how much the bill could be decreased by using more efficient models, or how much electricity use can be shifted to off-peak times.

The Energy Policy Act of 2005 (EPAct) implicitly recognised this flaw in electricity pricing, and encouraged utilities to provide time-based rate schedules reflecting variations during the day to all individual customers requesting it. However, the EPAct also said, ambiguously, that each state regulatory authority can decide whether to implement that provision. Correspondingly, Lawrence Berkeley National Laboratory estimates that only about 100 utilities (less than two percent) offered some sort of time-of-use rate for electricity customers in 2007.

Yet a preponderance of evidence suggests that pricing electricity more accurately will greatly improve the efficiency of the electricity industry, provide customers with proper price signals, and reduce wasteful energy use. One study provided residents with daily electricity prices for a month and found a 10.5 percent reduction in electricity use. Another analysis of residential electricity use from 1973 to 1980 found that "feedback" in the form of information detailing daily and weekly electricity prices reduced consumption between six and twenty percent. When Princeton University researchers gave residents of Twin Rivers, New Jersey, information about their level of electricity and natural gas use on a daily basis, consumption dropped ten to fifteen percent. Another study involved eight experiments tracking electricity use at 602 households over the course of many years. In some experiments, feedback was given three to four times a week, and in one experiment it was given continuously and informed households of the cost of their consumption every half hour. The researchers found that frequent, credible feedback about electricity prices resulted in 10-13 percent less electricity use than control groups.

In two of the metropolitan areas discussed above, New York and Atlanta, electric utilities have already started altering electricity prices to change behaviour. New York has experimented with "alternative rate designs" by offering time-of-use rates, day-ahead real time pricing, critical peak pricing, and pricing at real-time market rates. Researchers at LBNL surveyed 149 commercial and industrial customers in the Niagara Mohawk Power Corporation service area where the utility offered time-of-use tariffs for large customers with peak demand needs. They found that more than thirty percent of industrial customers responded by foregoing discretionary electricity usage and fifteen percent shifted usage from peak-periods to off-peak periods; forty-five percent of respondents installed demand reduction enabling technologies on site; and peak load for the utility was reduced by fifteen percent. Further south, Georgia Power introduced time-of-use metres for large industrial customers and, from 1992 to 2002, enrolled 1,650 customers to reduce peak demand by seventeen percent.

Because most people remain uninformed about the electricity-water nexus, a second form of feedback could be useful: making water usage associated with electricity generation "visible" by including it in people's electricity bills. California, for example, was the first state in the country to enact an Advance Recovery Fee on sales of some electronics. Whenever customers purchase new cell phones and televisions, a visible fee between six and ten dollars appears separately on receipts, showing consumers how much it will cost to collect and recycle some of the toxic elements inside their products. The same technique could be used in electricity billing, where consumers could be shown a separate line estimating the number of gallons of water used (and/or its associated cost) to produce the electricity they used within their home that day, week, or month.

The last type of information that utilities can offer consumers is education. The DOE noted in 2008 that eighty-eight percent of Americans would fail even a "basic" electricity literacy test. Another recent report in Kentucky found that forty-one percent of respondents identified coal, steel, and oil as "renewable resources." Because they remain uninformed and misinformed about electricity, many consumers will not understand the scope and scale of the electricity industry's water needs. Utilities could fund and promote electricity information and education campaigns to teach the public about electricity and water use. These could include grade-school classes on energy and the environment; public demonstrations and tours of renewable power facilities; free energy audits and training sessions for industrial, commercial, and residential electricity customers; and utility sponsored document "clearing houses" that offer websites, free books and articles to help consumers gather and process information in order to make more informed choices about their electricity use.

Designate National "Electricity- Water Crisis Areas"

Finally, each of the above efforts could be complemented by state or federal designation of national "Electricity-Water Crisis Areas." Cities and states expecting the most severe water shortages and thermoelectric capacity additions could form an interstate organisation to monitor and manage electricity-water resources, or could approach the federal government (most likely the FERC) to establish a National Electricity Water Policy Program Office through either Congressional legislation or an Executive Order. This Program Office could advise and assist local and state water managers and energy policymakers in their efforts to implement the efforts described above. The Program Office could chair a cabinet-level National Electricity Water Policy Interagency Task Force consisting of water and energy policy experts from the national laboratories, the DOE, the Department of Agriculture, the Department of Interior, the EPA, the Office of the Federal Environmental Executive, the Council on Environmental Quality, the Federal Energy Management Program, and the U.S. Geological Survey. This Task Force would have a few benefits over state-level action, since it could have the authority to coordinate and harmonize federal laws to stimulate the expedited implementation, permitting and siting of clean power facilities in all federal lands within or adjacent to National Electricity-Water Crisis Areas.

The pervasive water challenges confronting the electric utility industry in the United States suggest that local regulators, public utility commissioners, water managers, and electric utility operators start immediately responding to them. The water needs for expected future thermoelectric power plants - if they continue to use twenty-five gallons of water per kWh as they do today - in just eight metropolitan areas are so immense they could deplete the water available from Lake Lanier in Georgia and exacerbate interstate litigation between Tennessee, Alabama, and Florida by 2025. Biodiversity could perish along the Catawba-Wateree River Basin in North Carolina. Chicago could find itself embroiled in domestic and international legal disputes over the consumption and withdrawal of water from Lake Michigan. Households and businesses could run out of water from the South Platte River in Colorado. Rivers could stop recharging the groundwater needed for drinking and irrigation in Texas. Lake Mead and the Colorado River could continue to suffer drought in Nevada, inducing an agricultural crisis in California and Mexico. Fisheries along the Hudson River in New York could collapse. The Delta Smelt could become extinct in the San Joaquin River Basin in California. These impending but avertable risks serve an important reminder that climate change is not the only serious environmental issue facing the electricity industry.

Thankfully, lawmakers and electric utilities can implement six changes that would do much to avoid water shortages in the future. They could increase funding for research and development on new and alternative power plant cooling cycles, and place a moratorium on all new thermoelectric capacity additions, preserving the water available for other industrial, residential, and agricultural needs. They could aggressively implement energy efficiency and demand-side management programs to displace the need to construct new power plants. They could seriously deploy wind and solar power systems that use virtually no water when they generate electricity. (The combination of investments in energy efficiency, wind, and solar would not need to replace existing thermoelectric power plants, just offset new thermoelectric capacity additions.) Regulators could improve the feedback and information given to their electricity consumers about the electricity-water nexus and the water intensity of their own consumption habits. They could also form an interstate organisation or approach the federal government for assistance in designating National Electricity- Water Crisis Areas.

Interestingly, many of these actions would partially respond to a host of other challenges facing the industry unrelated to the use of water. The promotion of energy efficiency, wind, and solar would significantly reduce greenhouse gas emissions within the electricity sector, as all three technologies are less carbon intensive. A moratorium on future thermoelectric capacity additions would help diversify the fuels used to generate electricity, integrating more efficiency, solar, and wind into utility portfolios and serving as a hedge against fuel shortages and price volatility. Altering electricity prices would cut peak demand and create more informed consumers, an essential step towards any transition to a more decentralised electric utility sector involving small-scale distributed generation technologies. In this way, the proposals outlined here would not just save water and conserve electricity, but would also fight climate change, improve energy security, and educate electricity users. The benefits of such action surely extend beyond merely water and the electricity industry to society as a whole.

Chapter 7

Water Quality Regulations

The U.S. Environmental Protection Agency (EPA) estimates that agricultural pollution contributes to 60% of impaired streams, 30% of impaired lakes, 15% of the impaired estuaries, and 15% of the impaired coastal shoreline assessed. All told, more than 11.6 million acres of U.S. Rivers and lakes are considered impaired by excessive discharge of agricultural pollutants: soil, pesticides, pathogens, nitrogen, and phosphorus. U.S. policy makers have adopted carrotand-stick approaches to address some of the water quality problems linked to agricultural production and subsequent pollution. Federal funding targeted toward the mitigation of agricultural impacts on water quality has increased ("carrots"), and more stringent water quality regulations have been enacted pertaining to agricultural production ("sticks").

Specifically, funding for conservation practices on animal feeding operations (AFOs) and cropland through the Environmental Quality Incentives Program (EQIP) has been authorised to increase from 2002 levels of $200 million to more than $1 billion by 2005. EQIP provides agri-environmental payments to producers in order to generate broadly defined environmental benefits and to assist producers in complying with local, state, and federal water quality regulations. In addition, EPA has mandated nutrient standards for the largest AFOs, known as concentrated animal feeding operations (CAFOs). These standards essentially require manure nutrients generated on CAFOs to be spread on cropland at a rate no greater than the agronomic nutrient demand of the crops grown on that land, inclusive of commercial fertilizer applications. We couch these policy options in terms of agri-environmental "carrots" and regulatory "sticks," respectively.

The U.S. agricultural sector is likely to respond to these carrots and sticks in a variety of ways. A well-developed literature has examined the effects of agri-environmental payments for crop producers and their potential to reduce the environmental impacts of agricultural production. Recent national-level studies have also explored the implications of new water quality regulations for livestock and poultry production. These latter studies predict adverse economic impacts for affected AFOs, improved water quality, and increased commodity prices.

Figure : Water quality regulations and forms.

Notably missing from the literature are analyses of how these alternative approaches for improving water quality might interact across regions and across crop, livestock, and poultry sectors. Recent analyses have considered how agri-environmental payments might affect water quality markets or might interact with other conservation programs, such as Sodbuster. However, to the best of our knowledge, no previous study investigates the regional interaction of agri-environmental payments and water quality regulation in an endogenous animal and crop production setting.

For example, a corn producer in Iowa might receive agri-environmental payments in return for a reduced nitrogen fertilization regime. At the same time, a nearby swine CAFO might be willing to purchase the right to spread manure on that fanner's fields at the greatest extent allowable under a nutrient standard. In addition, because manure nutrients are not packaged as uniformly as commercial fertilizers, contain pathogens, and are generally more difficult to handle (Risse et al., 2001), crop producers may be reluctant to accept manure nutrients in lieu of commercial fertilizers. The willingness of producers to substitute manure nutrients for commercial fertilizers could play an important role in the effectiveness of carrot-and-stick approaches for achieving water quality improvements.

The potential interactions between and adjustments of crop and animal producers, given these sometimes competing carrot-and-stick incentives, could also generate secondary price impacts. In such cases, Berck and Huffman (2002) suggest a sector-wide assessment of economic adjustments. Moreover, because the impetus for these policies is to reduce adverse impacts on the environment from agricultural production, we conduct a regional and sector-wide assessment of potential economic and environmental implications.

Analysis findings suggest meeting nutrient standards would result in decreased levels of animal production, increased prices for livestock and poultry products, increased levels of crop production, and water quality improvements. However, estimated impacts are not homogeneous across regions or sectors. In regions with relatively less cropland per ton of manure produced, the impacts of agri-environmental policies are likely to be more pronounced. Impacts are generally smaller the more willing crop producers are to substitute manure nutrients for commercial fertilizers, and are generally larger the more animal feeding operations must meet nutrient standards.

Turning to the potential impacts on water quality, results of this analysis indicate surface water quality improves under the various scenarios considered. Overall, nitrogen discharge to ground and surface water falls by as much as 12.6%, phosphorus discharge falls by more than 30%, sheet and rill erosion falls by 6.7%, and pesticide discharge falls by 5%.

In several regions, however, there are unanticipated economic and water quality impacts. In some cases, agri-environmental payments to crop, livestock, and poultry producers may restrict animal production. Moreover, the discharge of some agricultural pollutants may increase under some carrotand-stick policies. Specifically, by requiring the spread of manure nutrients at no greater than agronomic rates, nitrogen leaching to groundwater may increase, and so may the discharge of sediment and pesticides to surface water in certain areas. Our results suggest that the use of agri-environmental payments to encourage the adoption of relatively benign crop production practices has the potential to offset many of these unexpected consequences.

The section below presents a regionalised sector model and illustrates the expected changes in agricultural production and prices given the imposition of land application constraints for manure nutrients in the presence of increasing agri-environmental payments. The next section describes the empirical analysis of carrot-and-stick approaches to managing water quality impairments from agricultural production. Simulation results are then presented for several scenarios, detailing the potential changes in market conditions, animal and crop sectors, and water quality. We conclude with a summary of findings and potential implications of key parameters for improving U.S. water quality.

Simulation Analysis

Those AFOS meeting nutrient standards spread the manure they generate on cropland at agronomic rates, or dispose of the manure in some other acceptable manner. When confined animal production within a region generates manure nutrients in excess of the assimilative capacity of the cropland, it can choose to find additional cropland for spreading, plant crops that consume more nutrients, raise animals that produce fewer manure nutrients, or reduce the number of animals produced.

Consider a national market for a representative livestock product with conventional supply and demand functions. The imposition of manure nutrient standards will result in increased costs of production for this product, reflecting nonlinear costs of manure management. In addition, due to increasing output prices for substitute goods throughout the animal sectors, demand shifts outward, establishing a new market-clearing quantity. The availability of agri-environmental payments further alters this market. Government payments lessen the supply contraction, and with similar responses throughout the livestock and poultry markets, the demand for this representative animal product will also shift.

We expect that agri-environmental payments will reduce the market displacement which would have occurred after the imposition of nutrient standards. With respect to higher food prices, consumers are better off when agri-environmental payments are available, but not as well off as when nutrient standards are absent. That said, consumers are also arguably better off when water quality increases. What is less discernable from this illustration is the extent to which animal producers benefit when agri-environmental payments are provided to AFOs who meet nutrient standards. Moreover, some regions will face greater costs than others when meeting nutrient standards, which will result in variable impacts across the United States.

There are many important parameters influencing agricultural sector and environmental responses to water quality regulation and agri-environmental payments. We constrain our analysis to three: nutrient standards (i.e., manure land application restrictions), agri-environmental payments, and manure substitution rates. We first consider a case when only CAFOs meet nutrient standards. For the purposes of this analysis, a CAFO is defined as an AFO with more than 1,000 animal units. These facilities represent 4.47% of the total number of AFOs in the United States, and will soon be required to meet nutrient standards. The quantity of manure generated by CAFOs exceeds 200 million tons, more than 46% of the total from confined animal operations.

Regional differences are notable. The percentages of CAFOs in the Southeast and Pacific regions are significantly higher than in other regions. And in the Northern Plains, Appalachia, Mountain, and Pacific regions, CAFOs generate more than 60% of the region's manure on confined animal operations.

As adoption of nutrient standards by all AFOs is the stated goal of the USDA, this study also considers the case when CAFOs and an additional 20% of the AFO manure nitrogen and phosphorus produced in a region meet nutrient standards.

This essentially reflects an increasing scope of the regulatory stick. Appalachia, Southeast, and Pacific regions have greater manure generation per acre of cropland than do other regions. Changes in economic performance throughout these regions could be the largest when nutrient standards are imposed due to the relatively high manure-to-cropland acre ratio. Greater environmental improvement in these regions might also be expected.

We next select a range of agri-environmental budgets to represent the carrot approach to inducing water quality improvements. To distribute payments to crop and animal producers based on EQIP provisions, it is assumed that 60% of the budget is allocated to livestock and poultry production to offset fixed and variable costs incurred by livestock and poultry producers when meeting nutrient standards. This includes manure nutrient testing, nutrient management plan development, and manure hauling costs. The remaining 40% of the budget is allocated to crop producers to encourage the adoption of best management practices on their cropland. These practices include residue management, conservation rotations, and reduced nitrogen fertilization. Three budget levels are examined: $0, $250 million, and $ 1 billion, to reflect funding levels indicative of authorised EQIP budgets through 2004.

Finally, we choose two manure-nutrient substitution rates over which to conduct the scenarios for this analysis. The willingness to substitute manure nutrients (substitute) is defined as the percentage of a region's agronomic demand for nitrogen and phosphorus (based upon crop requirements in that region) met by manure nutrients. Currently, 17% of corn producers and 8% of soybean producers supplement commercial fertilizer with manure as part of their crop fertilization regime [USDA/Economic Research Service (ERS), 2003]. While it is unclear to what extent substitution rates might change as AFOs adopt nutrient standards, it is realistic to assume this rate will increase, especially in regions facing binding nutrient standards. Increasing manure substitution could result from conservation programs such as EQIP or direct purchasing of spreading rights by livestock or poultry producers from crop producers. Therefore, substitution rates are allowed to vary between 20% and 30%.

Six scenarios illustrate the economic and environmental adjustments which may result from a carrot-and-stick approach to improve water quality in the United States. The baseline scenario (BASE) corresponds to the USDA forecast for crop and animal production in the year 2010, in the absence of nutrient standards or agri-environmental payments. Results are then presented from the case when crop producers meet 20% of their nutrient needs using manure generated on CAFOs (C20). The next two scenarios build on C20, by offering agri-environmental payments at the $250 million and $1 billion levels (C20-25 and C20-100), in line with EQIP funding expectations. Next, to reflect increased adoption of manure nutrient applications over time, crop producers are assumed to be willing to meet 30% of the nutrient needs using manure generated on CAFOs in the presence of a $1 billion agri-environmental budget (C30-100). Last, holding manure substitution constant at 30% and the budget at $1 billion, we assume an additional 20% of manure nutrients from previously unregulated AFOs are spread according to nutrient standards (AFO) to correspond to an increasing scope of the stick.

The Model

Aggregate agri-environmental payments for adopting environmentally benign crop (AEP_{rj}) and animal (AEP_{ri}) production practices are constrained by an exogenously determined budget (B), where B takes on values of $0, $250 million, and $1 billion. Crop producers are paid according to net environmental benefits generated from changing farm management practices. These benefits are broadly calculated to account for potential pollutant loading reductions to surface and groundwater (nutrients, pesticides, and sediments), reduced wind erosion, increased carbon sequestration, and increased soil productivity.

The model is solved iteratively for the imposition of nutrient standards and the use of agri-environmental payments. Because agri-environmental payments are not provided for land retirement under our assumptions, acreage responses occur in the first stage, where the imposition of nutrient standards is modelled. In the second stage, the model is re-evaluated in the presence of the agri-environmental payments, holding the acreage response constant. Agri-environmental payments to livestock and poultry producers are assumed to offset fixed and variable costs of nutrient standards.

The nutrient standards (1b) require the sura of each manure nutrient generated by animal production activity y within region to be less than or equal to the product of the regional substitution rate and agronomic crop nutrient demand, where f indexes nitrogen and phosphorus, respectively. Estimates of available manure nutrients by animal type are net the losses attributable to prevailing storage and handling technology. Agronomic demand is calculated using crop uptake values for nitrogen and phosphorus, accounting for losses due to denitrification, subsurface flow, runoff, and leaching. The affected AFO portion of available manure generation for each region and species is represented by that.

This constrained optimisation problem is simulated using the U.S. Regional Agricultural sector Model (USMP). The model accounts for production of major crop (corn, soybeans, sorghum, oats, barley, wheat, cotton, rice, hay, and silage) and confined animal (beef, dairy, swine, and poultry) categories, comprising approximately 75% of agronomic production and more than 90% of livestock and poultry production.

The USMP is a comparative-static, spatial, and market equilibrium model that incorporates agricultural commodity, supply, demand, environmental impacts, and policy measures. This model has been applied to various issues, such as climate change mitigation, water quality policy, and wetlands policy. The model permits the agricultural sector to adjust to the nutrient standards by substituting across production activities, and cropping and tillage practices with varying input requirements. This substitution is facilitated by nested constant elasticity of transformation functions allowing for interior solutions across activities and technologies.

Crop and animal production choices are linked to edge-of-field environmental variables using the Environmental Policy Integrated Climate Model (EPIC), which uses a daily time step to simulate weather, hydrology, soil temperature, erosion-sedimentation, nutrient cycling, tillage, crop management and growth, and pesticide movements to the field's edge. The transport of nutrients, pesticides, and sediment across the landscape is then calibrated to U.S. Geological Survey (USGS) estimates of regional pollutant loads.

Estimates of CAFO and AFO spreading practices on swine operations allow us to account for prior land application of manure in the simulations. Accordingly, CAFOs are assumed to spread manure on the nearest 155 acres, and the smaller AFOs are assumed to spread manure on the nearest 90 acres. While these numbers are not representative of the variety of animal operations across the United States, we argue that these values are reasonable for initial estimates of the environmental effects of excess manure utilisation at the Farm Production Region scale. The above levels provide a lower bound on the estimated benefits from meeting nutrient standards since many livestock facilities have little or no land on which to dispose manure. Given the acres currently receiving manure nutrients, we calculate the quantity of manure nutrient in excess of the crop requirements on those acres. These excess nutrients are available for potential leaching into ground waters and/or transport across the landscape into surface waters.

We obtain results portraying a potential range of national and regional changes in the U.S. agricultural sector following the application of agri-environmental carrots and regulatory sticks for water quality improvement by simulating various manure nutrient substitution rates for commercial fertilizers. The results suggest some of the costs of complying with nutrient standards will be passed along to consumers through higher retail meat, dairy, and poultry prices, with or without the presence of agri-environmental payments. In addition, changes in crop and animal production will vary regionally given pre-carrot and prestick production levels. Changing production patterns in agriculture will have subsequent impacts on regional water quality depending on underlying land and water characteristics.

Prices and Quantities

Under all scenarios, livestock and poultry prices increase and quantities decrease. The largest price changes occur in the poultry sector (e.g., 6.3% increase in the price of eggs), and the greatest production changes occur in the swine sector (e.g., 3.2% decrease in production). When agri-environmental payments increase, the individual sectors respond differently: amplifying the price and quantity changes in the poultry, dairy, and swine sectors, but muting the changes in the beef sector. When manure substitution rates increase, the impacts of nutrient standards on market conditions are lessened.

While price and production changes are examined in 10 major U.S. crop categories, our presentation is restricted to the largest two: corn and soybeans. The accompanying price and quantity changes for these crop sectors are not as large (less than 2% across all scenarios) nor as general as are those for the animal sectors. This is in part due to the dual role of cropland as a sink for manure nutrients and a source of feed grains for livestock and poultry operations. This sink role creates an incentive to plant crops that consume relatively high quantities of phosphorus (assuming the phosphorus constraint is the more limiting nutrient).

For example, note that the quantity of corn produced falls, as does its price under the C_20 and C20-25 scenarios. This market outcome occurs because the derived demand for corn as an ingredient in feed rations decreases more than the increase in demand for corn acreage as a means of disposal. The relatively large decrease in pork production, a major user of corn, lends support to this result. In addition, com production is associated with relatively high levels of sediment discharge, and increasing payments to crop producers in exchange for adopting environmentally benign production practices induces a movement away from corn rotations. This supply response to agri-environmental payments in turn lessens the decline in the price for corn. Conversely, even though the price of hay falls, production increases. One explanation for this result is that hay is a relatively high consumer of (i.e., sinks for) phosphorus, which outweighs the reduction in derived demand for hay as a source of feed.

Regional Responses

Undermost of the simulated scenarios, planted crop acreage declines marginally across the regions (less than 1%); however, cropland acres increase in the Southeast, Appalachia, and Pacific regions. These responses reflect an increase in demand for nutrient sinks in these regions, confirming our expectation that those regions with relatively high levels of manure generation per acre of cropland would experience the greatest changes in production. For regions in which livestock and poultry production increase under the C20 scenario, an initial increase in agri-environmental payments (C20-25) would further increase production. If agri-environmental payments further increased (a movement from C20-25 to C20-100), production would begin to fall. The opposite response occurs in those regions where animal production initially falls following the imposition of the initial nutrient standard (C_2O). It seems counterintuitive that providing agri-environmental payments to crop, livestock, and poultry producers, all else constant, would lead to lower animal production. One explanation for this result is that the agri-environmental payments to crop producers for the adoption of environmentally benign production systems might induce movement toward crop rotations not demanding large amounts of phosphorus and nitrogen. Movement to these alternative crop mixes would serve to make the nutrient application standards more difficult to meet for those affected AFOs, essentially decreasing the availability of manure nutrient sinks. However, at higher levels of agri-environmental payments, crop producers appear to be adopting management practices, such as residue management, which do not focus as much on crop rotations. Hence, the production responses are marginal.

Sector Responses

The changes in national-level prices and quantities translate into differing regional responses as expected. Corresponding adjustments in net returns are not as straightforward. At lower manure substitution rates, the price impacts are sufficient to compensate aggregate decreases in livestock and poultry production. Nationally, net returns increase with increasing agri-environmental payments. When manure substitution rates are at 30%, the price effect no longer dominates the production (and carrot) effect, and national net returns to animal production fall. However, by increasing the scope of the stick (the AFO scenario), the price effect in conjunction with higher levels of agri-environmental payments results in increasing net returns (over the BASE scenario) for the livestock and poultry sectors.

The crop sector results are nearly opposite of the livestock and poultry sector results. Initially, aggregate net returns fall due to the decrease in prices, which are relatively larger than the increase in total acreage planted. However, as agri-environmental payments increase or as manure substitution rates increase, the change in crop returns becomes positive. Note that, under our assumptions, the agricultural sector as a whole might experience increases in net returns by amounts greater than the agri-environmental budget.

This finding does not suggest all sectors or regions will share in these increased returns. Taking a closer look at the 10 regions, we observe that in many instances net returns fall as the carrot grows and the scope of the stick widens. Falling net returns for livestock and poultry production are especially evident in the Southeast, Appalachia, and Pacific regions, and do not seem to be affected by increasing availability of agri-environmental payments. Actually, initial offerings of agri-environmental payments appear to lead to marginally lower net returns-indicating, in these regions, the agri-environmental incentives to produce crops using systems that consume relatively fewer nutrients are greater than the incentives to provide animal producers with nitrogen and phosphorus sinks. Even at high levels of agri-environmental payments, net returns to animal production in the Southeast continue to fall.

Based on these results, a question is raised about the overall impact of agri-environmental payments on net returns when prices are allowed to adjust. The transfer efficiency for animal production is less than 60% under the various scenarios, implying that even though animal producers are assisted in complying with nutrient standards by agri-environmental payments, the subsequent price effects mute the impact of carrots on net returns. Initially, the transfer efficiency for livestock and poultry producers is only 33%. Specifically, under C20-25, a $150 million (i.e., 60% of $250 million) transfer to livestock and poultry producers to assist in compliance results in an increase in net returns of $50 million. This transfer efficiency rises to 51% under C20-100, when an additional $450 million in agri-environmental payments are allocated to livestock and poultry producers.

Conversely, the transfer efficiency of agri-environmental payments to crop producers is much higher than for animal producers incorporating price adjustments, exceeding 100% in our policy scenarios. Under C20-25, a transfer of $ 100 million to crop producers results in a more than fourfold increase in net returns (i.e., from - $596.85 million to - $124.86 million). An additional transfer of $300 million in agri-environmental payments to crop producers under C20-100 results in a gain of $573.15 in net returns, a transfer efficiency of 191%.

In looking at the national totals for both crop and animal sectors, a $250 million investment in our agri-environmental program yields an increase in net revenues of more than $700 million when coupled with the regulatory stick at the lower manure substitution rates. These estimates include the potential savings in commercial fertilizer costs, potential costs incurred in meeting nutrient standards and in providing agri-environmental benefits, potential impacts of price and production changes, and transfer effects of agri-environmental payments. At the higher budget, national net returns increase by more than $1.7 billion for agriculture as a whole. At higher manure substitution rates and when more AFOs meet nutrient standards (i.e., C30-100 and AFO), increases in net returns are not as high as under the C20-100 scenario, yet they still exceed $1 billion.

Environmental Impacts

The use of EPIC allows us to examine the environmental implications resulting from our carrot-and-stick scenarios. In particular, we estimate the potential quantity of nitrogen discharged into surface and groundwater, and the potential quantities of three additional contaminants discharged into surface water-phosphorus, sediment, and pesticides.

The potential changes in the discharge of nitrogen to surface waters and leaching of nitrogen to groundwater reveal some unintended effects of our carrot-and-stick scenarios for improving water quality. Across all regions and scenarios, the amount of nitrogen discharged to surface waters falls from the pre-carrot and stick scenario. Nationally, these reductions range from 10% to 16%. Increasing the scope of the stick and the size of the carrot leads to larger reductions, although greater manure substitution rates mute this response.

Nitrogen leaching increases under the C20 scenario, but falls with increasing agri-environmental payments and increasing manure substitution. A broadening of the scope of the stick from the BASE scenario to C20, or from C30-100 to AFO, results in a more binding nutrient standard, creating an incentive to expand cropland acres and adjust acreage share in favour of crops that consume relatively more nutrients and leach more nitrogen, particularly in the Southeast, Appalachia, and Pacific regions. Because the phosphorus constraint is more binding than the nitrogen constraint, crop producers will have to supplement the new cropping patterns with additional commercial nitrogen fertilizer, which serves to undermine the reductions in manure production.

For example, a closer examination of the potential adjustments occurring in the Pacific region reveals a general expansion in crop production across the scenarios, especially in corn and hay production. Both of these crops exhibit relatively high levels of nitrogen leaching. Furthermore, in areas of California, the potential expansion in cropland and in cotton, rice, and barley production results in elevated levels of nitrogen leaching. Nevertheless, the nitrogen prevented from reaching surface waters is of a greater magnitude than the relatively small increases in nitrogen leaching across all regions and scenarios. Overall, the reduction in nitrogen discharged to ground and surface waters ranges from 6.2% to 12.6%.

As for other measures of potential water quality impairment, changes in phosphorus loading are observed to follow the same pattern as nitrogen discharged to surface water, with reductions in phosphorus ranging from 24.8% to 37.6%. The results indicate that increasing the scope of the stick, the size of the carrot, and the manure substitution rates all contribute to reduced phosphorus discharge. However, as with nitrogen leaching, the quantities of sediment and pesticides discharged into surface waters increase in the absence of agri-environmental payments. As agri-environmental payments increase, loadings of sediment and pesticides fall. The change in soil erosion ranges from -0.7% to 6.7%, and pesticide-loading change ranges from-0.9% to 5.5%.

Soil erosion is greatest at the lower manure substitution rate under the smaller stick (C20-100). That is, by relaxing the constraints on spreading manure nutrients (moving from C20-100 to C30-100), less land leaves production in regions with decreased crop production (e.g., the Corn Belt) and less land comes into production in regions with increased crop production (e.g., the Pacific). The relative changes in soil erosion rates across these regions would result in an additional 330,000 tons of sediment being discharged under C30-100 relative to C20-100. The amount of pesticide discharged to surface waters declines with increasing agri-environmental payments, with increasing acceptance of manure nutrients, and with an increasing scope of manure nutrient standards for AFOs. These reductions range between -0.9% under the C20 scenario and 5.5% under the AFO scenario.

Agricultural Sector Analysis

A number of efforts at the local, state, and federal levels aim to reduce potentially adverse impacts of agricultural production on the environment in general and on water quality in particular. Some trends that are illustrative of these efforts include carrot approaches (the increased level of support in recent Farm Bill legislation for crop, livestock, and poultry producers to implement environmentally benign production practices) and stick approaches (the recently promulgated rules for manure nutrients generated on CAFOs).

This study analyses potential economic and environmental implications of these carrot-and-stick approaches to improving the quality of U. S. water resources. In addition, we consider how the willingness of crop producers to substitute manure nutrients for commercial fertilizers might influence the changes brought about by carrot and stick policies. These parameters form the basis of six potential scenarios depicting how agri-environmental payments, the scope of manure nutrient standards, and manure substitution rates might evolve in the United States.

First, we compare results when no payments or nutrient standards exist with the scenario where crop producers are willing to meet 20% of their crop nutrient demand using manure (BASE to C20). Building on this scenario, the effect of providing agri-environmental payments to crop and livestock producers is evaluated, somewhat akin to the

Environmental Quality Incentives Program. By increasing the budget from $0 to $250 million to $1 billion, the carrot effect is investigated (scenarios C20 to C20-25 to C20-100). Two additional scenarios are then examined to see how increasing manure substitution rates and coverage of nutrient standards might differ from our earlier results (C30-100 and AFO).

A wealth of regional- and sector-level results emerge from these scenarios, illustrating how agri-environmental policies might affect agricultural production and improve water quality in the United States. In general, carrots and sticks result in decreasing levels of animal production, increasing levels of crop production, and increasing prices for livestock and poultry products. In particular, poultry and dairy products could see substantial price increases when the willingness of crop producers to substitute manure nutrients remains low. Nevertheless, adverse impacts on net returns to both crop and animal producers can be mitigated by providing increasing agri-environmental payments.

For example, without agri-environmental payments (C20), six often regions experience decreasing net returns to crop production; however, by including $400 million in agri-environmental payments for crop producers (C20-100), all regions experience increasing net returns to crop production. For those regions that experience adverse impacts on net returns to animal production (the Southeast, Appalachia, and Pacific), it appears the willingness of crop producers to substitute manure nutrients has the most bearing on reducing these losses. Because net returns for crop producers also increase with increasing manure substitution, use of carrots to induce crop producers to utilise more manure nutrients may represent an avenue for future inquiry.

Results indicate agri-environmental payments can offset some unintended consequences of the nutrient standards on water quality. Specifically, by requiring certain AFOs to spread manure nutrients at no greater than agronomic rates, there is potential in some regions to increase both nitrogen leaching to groundwater and increase discharge of sediment and pesticides to surface water due to changing crop levels and composition. However, when agri-environmental payments to crop producers are used to encourage the adoption of environmentally benign production practices, the subsequent reductions in cropland loadings offset any potential increases induced by the nutrient standards. Overall, nitrogen discharge to ground and surface water might fall by as much as 12.6%, phosphorus discharge by more than 30%, sheet and rill erosion by 6.7%, and pesticide discharge to surface waters by more than 5%. While it is beyond the scope of this chapter to place a value on the benefits of these reductions, findings suggest coordinating agri-environmental payments and manure spreading regulations can significantly enhance water quality in the United States.

Aggregate analysis of the U.S. agricultural sector cannot reveal how individual operations would benefit or suffer from these trends. However, if only the largest animal feeding operations meet nutrient constraints, the costs of compliance would fall on CAFOs and the benefits from secondary price effects will accrue to smaller AFOs. When crop producers' substitution rates for manure nutrients remain at or near current levels (i.e., a 20% substitution rate), the secondary price effects are sufficient to compensate much of the livestock and poultry sectors for the costs of meeting nutrient standards. However, at higher manure substitution rates, the increased costs of transporting manure, manure testing, soil testing, and developing a manure management plan outweigh compensating price effects and foregone commercial fertilizer purchases, resulting in reduced net returns for the livestock and crop sectors. Also, at higher manure substitution rates, consumers benefit from lower price increases and greater increases in water quality.

Ecological Islands and Vickers Focus Adaptive Transitions in the Pre-contact Plains

Ongoing research in southwestern Manitoba is addressing archaeological sites associated with the hitherto sparsely represented Vickers Focus. These sites are important because they document part of a much larger phenomenon of pre-contact population movement into the eastern Plains. Throughout much of the eastern Plains this involved either semisedentary village horticultural expansion (i.e. Plains Village) or movement of forager populations (i.e. Plains Woodland). In southwestern Manitoba, sites associated with this phenomenon are tightly associated with isolated forest and wetland zones surrounded by mixed grass prairie. The biotic character of these ecological islands was extremely important for the Vickers Focus, and is useful for understanding the process of behavioural adaptation to new environmental circumstances. Vickers Focus sites are usually found removed from the primary river systems, and are associated with localised wetlands.

Until recently, it was thought that these sites occurred singly in widely scattered localities across the Manitoba Escarpment. However one area has yielded a dense cluster of large Vickers Focus sites within the Souris River Basin. We have named this locality Makotchi-Ded Dontipi, a Dakota phrase meaning "the place where we live." Vickers Focus is characterised, in part, by its stylistically diverse ceramic assemblage, relatively large site size, and unconventional site placement. The ceramic assemblages include ware types generally associated with Plains Village and Late Plains Woodland societies within the Missouri and upper Mississippi River basins.

Considerable ambiguity exists in the use of the terms Plains Woodland and Plains Village. We follow the perspective offered by Gregg (1994) whereby these terms denote "cultural traditions" or general adaptive strategies. In the northeastern Plains, the term Plains Village is commonly used to indicate large, often fortified, earth lodge villages that were sustained by foraging and horticulture. Such communities are most commonly found in the Middle Missouri River valley, although regions beyond the Missouri Trench have yielded sites with at least some of these traits. The term Plains Woodland is used to imply a more generalised and mobile forager economy employed by groups who produced pottery, constructed burial mounds and evince other influences from the woodlands east of the Mississippi River. This is generally interpreted as indicating population movement into a new habitat which began as early as 1,500 years ago.

In addition to population movement into the northern Plains, no doubt diffusion of new ideas and technology also figures in the phenomenon. We use the term Plains Woodland to reflect an ongoing process of population movement out of the Eastern Woodlands that persisted up to the post-contact era. In sum, we use the terms Plains Village and Plains Woodland to imply the end points of a continuum of economic and subsistence options. The former implies some level of commitment to horticulture and sedentary village life, while the latter suggests reliance upon intensive foraging and greater degrees of population mobility. An unresolved question is where along this continuum one shifts from using one descriptor to the other.

The closest cultural associations for the Vickers Focus are Sandy Lake, Oneota, Initial Middle Missouri, and the Scattered Village

Complex. In the Mississippi River basin, such societies possessed mixed economies relying to varying degrees upon horticulture and foraging, and sometimes with a village-based settlement system along streams draining the eastern Great Plains. To date, only one Canadian archaeological site has yielded conclusive evidence of horticultural activities. The Lockport Site, located on the Red River in southeastern Manitoba, yielded bell-shaped storage pits, Sandy Lake and Oneota-like ceramics, and radiocarbon dated scapula hoes and maize kernels. We do not, however, propose that the horticultural component at Lockport is a satellite Middle Missouri community. Just where this important site fits in the cultural taxonomy must await the publication of a comprehensive site report. The relevant Lockport component yielded five calibrated radiocarbon dates ranging between AD 1200 and AD 1500. Vickers Focus calibrated radiocarbon dates fall between AD 1350 and 1650. They overlap and post-date the horticultural component at Lockport, and coincide with the onset of the cooler climatic conditions of the neoboreal episode.

It remains to be determined whether Vickers Focus sites located on the Manitoba Escarpment of southwestern Manitoba reflect Plains Village cultural influences including horticultural production. Alternatively, perhaps they document a Plains Woodland economy based on seasonally scheduled intensive foraging. To date, no Vickers Focus components have yielded direct evidence of horticultural production, storage features, structures or fortifications. While this negative evidence indicates a Plains Woodland adaptation, current interpretations may be a function of small sample size and inappropriate excavation methodologies. That is, our exploratory research has focused upon shovel test pits, dispersed one metre square excavation units, and only minimal microbotanical and macrobotanical analysis. On the Manitoba Escarpment the strongest Plains Village technological influence is apparent at the Lowton Site, located north of the Pembina River Trench. This site was largely destroyed by cultivation and has been heavily surface collected. Archaeological materials include a wide range of ceramic wares associated with the upper Mississippi River basin, assorted stone hoes, axes and mauls, and catlinite disks and tubular pipe fragments (Nicholson and Malainey 1995).

Calibrated radiocarbon dates indicate that Lockport, Lowton, Lovstrom and the Makotchi-Ded Dontipi sites form a weakly expressed east to west temporal continuum. The technological evidence of horticulture and other Plains Village traits decline through time and as one moves west. If this pattern is valid, then it might suggest culture change among groups of generic Plains Woodland/ Plains Village origin that, over the course of a few centuries, expanded into the northeastern plains, and then gradually abandoned the semisedentary horticultural economy. In its place, they employed an economy and settlement pattern increasingly focused on intensive foraging and bison hunting. We suspect that a logical endpoint of this transition may be epitomised by a conventional northern Plains mobile bison hunting economy. Perhaps further refinement of this model may aid in the interpretation of the cultural origin of such entities as the "Middle Missouri" influenced Mortlach culture. This speculation is consistent with the recent discovery of late pre-contact and protohistoric Mortlach components within Makotchi-Ded Dontipi, and observations by Vickers (1994:25-26) regarding an apparent east to west geographic expansion of Mortlach.

These observations are not meant to imply strong cultural linkages between the occupants of the Lockport site, Makotchi-Ded Dontipi, and beyond to Mortlach occupation of the northern Plains. Rather, we propose that among the autonomous groups expanding out of the Mississippi River headwaters after 1200 AD, there was a pattern of gradual abandonment of Plains Village-Woodland economies in favour of a Plains-oriented one heavily reliant upon bison hunting. While dependent upon bison hunting, Vickers Focus sites are significantly different from those associated with roughly contemporaneous mobile bison hunters such as Blackduck and Mortlach. With the exception of the large, repeatedly used, Stott Site (predominately Blackduck), most such sites are comparatively small, and consistent with a highly mobile lifestyle. With its tight temporal and spatial clustering of sites, Vickers Focus occupation within Makotchi-Ded Dontipi reflects a unique settlement pattern focused on forested wetlands surrounded by mixed grass prairie.

The Environmental Context

The research area is located within the glacial Lake Hind basin. This glacio-lacustrine plain developed approximately 12,000 years ago when meltwater became trapped along the southwestern flank of the Laurentide ice front to form glacial lakes Souris and Hind. The Lake Hind basin consists of fine silts and sands, with a series of relict fluvial deltas along its flanks and within its basin. The Souris River basin is composed of impervious sedimentary and bedrock deposits mantled with well-drained sand and silt. These impervious deposits trap and retain precipitation in an internal drainage basin, forming the Oak Lake Aquifer. Throughout the Holocene Period, ground water accumulated within the aquifer to the point that surface undulations filled to create small lakes and sloughs surrounded by arboreal vegetation. While the topography is quite subtle, some areas possess dramatic local relief caused by aeolian reworking of deltaic sediments. Prevailing winds eroded and transported unstable sandy sediments. When intercepted by the wetland tree margins, these winds lost velocity, and dropped their sediment load to form sand dunes along the windward sides of the wetlands. The resultant landscape consists of stabilised sand dunes interspersed with wetlands of various sizes. This complex topographic and hydrological situation favoured the development of mixed forest, wetland and meadow "islands" surrounded by mixed grass prairie. For the purposes of this chapter it is sufficient to observe that:

1) as the wetlands ultimately derive from the Oak Lake Aquifer, they were comparatively stable and drought resistant;
2) these wetlands sustained concentric rings of aquatic and arboreal vegetation surrounded by mixed grass prairie;
3) the woodland/wetland habitats sustained a wide range of fish, waterfowl and mammals that otherwise would not have been locally available;
4) the wetland forest fringes acted as localised windbreaks, thereby fostering the development of linear sand dunes which flank and further protect the wetland habitats;
5) the configuration and density of the wetlands reduced the frequency and severity of prairie fires which otherwise would have destroyed the woody vegetation; and
6) the abundance of sedges and grasses within this sheltered microhabitat provided ideal winter range for bison in addition to the woodland species.

The net effect was the formation of an isolated and relatively stable forest-wetland ecological niche surrounded by mixed grass prairie. The historic nature and significance of this ecological island is presently obscured by the local extinction of bison, modern agricultural settlement, the suppression of chronic prairie fires, and recent development of dense aspen-oak forest on uncultivated lands. Also important is the dramatic decline in ground water levels within the Oak Lake Aquifer. The ecological implications of these transformations are readily apparent upon comparison of historic and modern vegetation maps. Ernest Thompson Seton's 1905 vegetation map offers the best indication of the prehomestead era biomass throughout southern Manitoba. Twenty-five years after the beginning of homestead settlement, he reported that grassland still dominated much of the region. Significantly, he also recorded the distribution of isolated forest habitats:

1) upon uplands within the Manitoba Escarpment and upon high end moraines;
2) within deeply incised and sheltered river valleys and lake shores; and
3) on lands underlain by the Oak Lake and Assiniboine aquifers.

This vegetation pattern forms a sharp contrast to the modern situation with the dramatic southward expansion of the aspen-oak parkland.

Seton's map has limited utility for understanding the ecological details of the Oak Lake Aquifer region because of its small scale and coarse resolution. To generate maps suitable for addressing Vickers Focus settlement, we have relied upon the late 19th century records of the Canadian Dominion Land Survey. These field notes and township plans report relief, vegetation cover, sediment quality, and surface hydrology along the "road allowance" boundaries of each square mile section. When compiled into a composite map, these archival records offer a relatively comprehensive pre-homestead vegetation inventory. We argue that since this composite map predates the early 20th century landscape perturbations, it can be used to model the ecological situation in the late pre-contact period.

These archival sources are being verified by archaeological reconnaissance and landscape mapping using a differential Global Positioning System (dGPS). Our reconnaissance has revealed at least 11 archaeological sites. The boundaries and constituent subareas of five major sites have been defined by shovel test transects and test excavation. Radiocarbon dates have been generated from five sites, and six sites have also been sufficiently mapped to permit landscape reconstruction and modelling. Vickers Focus sites are rare in the Manitoba archaeological site registry. Until recently only three sites widely scattered across the Manitoba Escarpment (Lowton, Lovestrom and Jonas sites) had been assigned to the Vickers Focus. The last five years of reconnaissance within the Oak Lake region has tripled the provincial inventory.

Most of these newly discovered sites are within a 4 square km area along the western side of the Lauder Sand Hills. This is a conservative estimate given the preliminary nature of our exploration in this densely forested area. We have not surveyed the main body of the Lauder Sand Hills, an area of forested stabilised sand hills and wetlands that encompasses about 4,400 hectares.

Recent road construction in the Rutledge Sand Hills north of Oak Lake has revealed the Taylor Site, yet another large Vickers Focus site. The Rutledge Sand Hills encompass 2,600 hectares and are remarkably similar to the Lauder Sand Hills. The accidental discovery of the Taylor site in this zone demonstrates that suitable wetland-forest microhabitats throughout the Oak Lake Aquifer region attracted Vickers Focus settlement. As we have not begun investigation in this latter area, we can only speculate about the nature and density of sites within it.

Nicholson has proposed that horticulture might have supplemented hunting and gathering as the basis of the Vickers Focus economy, particularly at sites located in southcentral Manitoba (i.e. Lowton, and Lovstrom). As noted earlier, this is circumstantially indicated by the recovery of possible stone hoes, and a range of ceramic wares reminiscent of the Mississippi River headwaters. Furthermore, virtually all Vickers Focus sites share an association with warm sandy sediments and a high water table. In the northeastern Plains, successful horticultural production is largely dependent upon sufficient frost-free days and high daily temperatures. Also important is a steady supply of water to withstand midsummer drought and to protect garden plots from prairie fires. We argue that these criteria are met by the sand hill-wetland mosaic associated with the Oak Lake Aquifer.

While this possibility is logical, our preliminary reconnaissance and test excavations have yet to yield evidence supportive of horticultural production. As stated, this might be due to sample bias and the preliminary stage of sedimentological and archaeobotanical analysis. Parallel with the ongoing search for evidence of horticulture, we have begun developing alternative explanations for the tightly focused site distribution within these forest-wetland microhabitats.

Paleo-Environmental Context

Virtually all of the wetlands within the Lauder Sand Hills have disappeared as a consequence of dramatic groundwater declines within the Oak Lake Aquifer. These former wetlands consisted of sedge-willow wetlands or small pothole lakes surrounded by aspen groves and mixed grass prairie. The wetlands supported stands of willow, river birch and other water-loving shrubs and trees. However, in 1969 a drainage canal was excavated from Maple Lakes to the Souris River. This drainage canal inadvertently transformed the local drainage patterns, and sharply accelerated rates of soil desiccation that began after 1880. The declining ground water level had catastrophic effects upon aquatic plants growing in these sandy well-drained soils. The aquatic vegetation was immediately killed off. The willows also died because their root systems were unable to reach the new water table level before the foliage sustained massive die-back.The dried potholes and surrounding willow thickets were quickly colonised by aspen and are now completely overgrown. This rapid development of aspen-oak forest is consistent with a general pattern of forest development on uncultivated lands throughout southern Manitoba. While the area is now heavily overgrown, intensive surface reconnaissance reveals dead aquatic trees in localised pockets throughout the area. By mapping this dead vegetation in conjunction with other landscape characteristics, we can estimate the distribution of former wetlands. These data confirm the wetland distribution reported in the 1880 DLS maps and document unreported wetlands located beyond the road allowances.

Archaeological reconnaissance demonstrates that virtually all presently known Vickers Focus sites are located either:

1) adjacent to isolated former wetlands;
2) upon gently rolling uplands adjacent to extensive wetlands;
3) within small meadows flanked by sinuous sand dunes; or
4) on the lee side of sand dune features protected from summer or winter prevailing winds.

Shovel test pits in the sandy uplands consistently yield few or no archaeological materials, thereby strengthening the spatial association of sites with former wetlands. These strong spatial correlations indicate that the relict wetland vegetation is a powerful proxy variable that indicates the presence of former wetlands and archaeological sites. These spatial associations offer a useful technique for predicting Vickers Focus settlement, and enable interpretation of economy and settlement pattern. While such proxy variables are powerful predictors, they do not necessarily explain site function.

This is illustrated by the Jackson Site. This extensive site is predominately Vickers Focus, with a small, localised earlier Blackduck occupation at its south end. It has been mapped at a 25 cm contour interval and is at least 26,000 square metres in extent. The topographic map has been supplemented with a preliminary vegetation inventory, systematic shovel testing, a proton magnetometre survey, and strategic test excavation. These data indicate gently undulating topography with a sinuous sand dune defining the southern limit of human occupation. Shovel test transects reveal dense concentrations of archaeological materials within an area of shallow hollows and localised knolls in a dense aspen grove.

Air photos from the late 1940s indicate that the dense forest cover in the area is a recent phenomenon. During the early 20th century a much more open grassland-forest mosaic existed, with smaller, more isolated forest groves. Several forest groves observed in the 1940s air photos have been inspected and mapped using a differential GPS system. This demonstrates the rapid spread of these groves, in some areas forming contiguous forest. It has also consistently revealed dead willow trees within localised hollows at the core of the forest groves, indicating that they developed around nowdry wetlands. These hollows are clearly former wetlands where groundwater from the Oak Lake Aquifer reached near the surface. While we have documented significant declines in the water table within 30 years of initial homestead settlement, sufficient groundwater remained to sustain these localised wetlands until after the 1969 construction of the Maple Lakes drainage canal. According to local informants, after this date many localised wetlands and humid meadows dried up and were colonised by aspen. We expect that several currently undocumented Vickers Focus sites are associated with the many isolated forest-wetland microhabitats that formerly existed in the area.

The Jackson Site is quite distant from the major wetlands reported on the 1880 legal survey maps, however, dead willow thickets are found in localised hollows within the site, indicating the former hydrological situation. As part of an experiment in landscape modelling, the topographic information was transferred into a Geographic Information System (GIS), and rendered as a continuous relief "Digital Elevation Model" (DEM)'. The spatial distribution of dead willow was then plotted and treated as indicative of the margins of former wetlands. The densest distribution of willows forms a partial concentric ring around the margins of a flat bottomed meadow (Locality Z).

These dead willows were used to identify and constrain zones that might be former wetlands. All elevations below 435.75 metres ASL were treated as former wetlands, and land up to 25 cm above this level was reclassified as the aquatic vegetation fringe surrounding the wetlands. The spatial distribution of the "virtual wetlands" coincides with most stands of dead willow observed at the Jackson Site. This is consistent with observations deriving from sediment test pits within Locality Z. These tests reveal sand richly stained with iron compounds within 5 cm of the surface. These so-called "flame structures" developed where mineral-rich ground water underwent transevapouration, leaving behind iron oxide precipitants. The location of these red-stained sediments suggests ground water at or very near the surface.

As part of further experimentation with vegetation modelling, we expanded the hypothetical vegetation buffers to propose the spatial extent of aspen-oak forest groves surrounding these wetlands. In this case, we merely constrained the woodland buffer to between 436 and 436.25 ASL. The resultant map reveals a series of closely spaced small wetlands, each associated with concentric rings of woody vegetation. As an aside, further exploration of Jackson and other sites in the area indicates that relief and relict vegetation are insufficient to model former hydrology and vegetation cover. One must also consider the existence of subsurface sedimentary unconformities (such as impervious gravel and clay lenses) that influence local ground water drainage patterns. Our provisional modelling efforts are sufficient to demonstrate the spatial configuration of these wetlands, and that the vegetation buffers likely coalesced to form a relatively contiguous aspen grove that was somewhat protected from prairie fires. Within this small pocket of wetlands and forest, patterned accumulations of cultural material have been discovered in shovel test pits and one metre square excavations. The nature and pattern of these vegetation reconstructions significantly clarify interpretations of intrasite artifact patterning at the Jackson Site.

These site modelling interpretations are consistent with our initial generalisation that Vickers Focus sites are tightly associated with forested wetlands and warm sandy soils. Given our expectations, this might indicate that Jackson is a good candidate to intensify the search for evidence of horticultural production. If small-scale horticultural production occurred at the Jackson Site, then it is reasonable to have expected some concentration of archaeological material around the largest and best defined wetland. The most probable zone for wetland fringe garden plots is associated with Locality Z. Our experience excavating at this site in summer demonstrates that the southfacing knolls and dune slopes immediately north of Locality Z enjoy maximum sunshine and the highest daily temperature. This is a function of the lack of forest cover on the well-drained local uplands and the south-facing aspect. In the context of short growing seasons, we expect that the well-watered margins of Locality Z would offer optimal conditions for horticultural production, and that we should expect disturbed soil profiles and significant archaeological recoveries. This was not borne out by the shovel testing. The vast majority of materials are concentrated in the northern two thirds of the site.

The shovel test transects across the site area revealed a tightly constrained zone of minimally processed bison bone located within a small willow-choked hollow at the north end of the site. This site subarea has been interpreted as a small bison entrapment and kills area. The bone bed also contained stone projectile points and small rejuvenation flakes consistent with repeated re-sharpening of butchering tools. To the south, dense concentrations of severely smashed bison bones were encountered within and immediately south of a large hollow containing dead willow trees. These deposits indicate a waste disposal zone after intensive bone grease extraction activities. The most diverse mix of artifact classes (including lithics, fire-cracked rock and ceramics) was encountered upon a localised knoll surrounded by former wetlands. We propose that this knoll area contained the primary domestic habitation area, with the willow bog to the west serving as a convenient midden area. The "habitation" zone would have been particularly attractive as it was well drained and protected by the wetland fringe forest groves that coalesced there. This offered shelter from the summer heat, and shelter and firewood for winter occupation. It remains to be conclusively addressed how and in what season bison were captured and killed at the Jackson Site. Preliminary faunal analysis suggests a winter entrapment event. Provisional interpretation of the nature and disposition of the small kill zone is heavily dependent upon the hydrological and vegetative model offered above. The kill zone is limited to a localised hollow containing a dense stand of dead willow. This hollow drains two shallow swales located to the north and northwest. Beyond and between these swales are sandy knolls that extend to the northwest as gently undulating prairie dotted with small aspen groves. Five hundred years ago the north end of the Jackson Site probably was fully exposed to the winter prevailing winds. The forest fringe along the northwestern side of the site would have acted as a wind break that accumulated drifting snow. The dense willow thickets in the most northerly wetland hollow would have been particularly deeply drifted over. By contrast, the exposed rolling upland prairies to the north and west likely had minimal snow accumulation. Inspection of the Jackson Site in the late winter of 1996-97 confirmed that deeper accumulations of snow occurred in the kill zone compared to the adjacent exposed uplands. This differential accumulation occurred in spite of the significantly expanded forest cover of the past 50 years.

If we accept the vegetation reconstruction, then driving bison from the open prairie and into snow-packed willow thickets within deceptively sheltering forest groves would have been an effective means of entrapment of small groups of bison. The entrapment capability of these drifted-over hollows could be enhanced by weaving brush and poles into the thickets to create an impenetrable barrier.

Given the tight temporal overlap of calibrated radiocarbon dates the Jackson bison kill and domestic habitation subareas may represent either a single event or perhaps closely spaced occupations. The successful bison entrapment may have enabled sustained winter occupation. In this case perhaps the spatial association of archaeological deposits with the wetland margins is due to their utility for communal bison entrapment, and because it offered localised winter shelter and firewood.

Environmental contextualisation of the Jackson Site greatly expedites its interpretation. If our interpretations of site function and seasonality are correct, then continued excavation in such highly productive sites may be counterproductive for the investigation of the role of horticulture in the Vickers Focus economy. While the Jackson Site clearly represents an intensively occupied site, the economic basis of occupation is not clear-cut. We initially proposed that the sedimentological and hydrological correlations implied optimal summer conditions for small-scale horticultural production. While this might prove appropriate for some sites, the intrasite artifact patterns and our preliminary interpretations of seasonality indicate otherwise.

The Vera, Schuddemat and Twin Fawns sites are reminiscent of Jackson in terms of landscape characteristics and comparatively high bison bone densities. Such sites indicate that intensive bison procurement was an established focal element of Vickers Focus settlement, at least on a seasonal basis. However, sites such as Bradshaw, Jocelyn, VBD, Hollow B and Louttit share a different landscape context and artifact distribution pattern. Preliminary subsurface investigation of these sites reveals significantly less bison bone, and much more diffuse and ephemeral artifact distributions. These sites are either very localised within small hollows, or reflect discontinuous clusters of material culture within larger meadows. These localities are sheltered by dunes and are in close proximity to ground water as indicated by the proxy variables. At least the Bradshaw Site faunal recoveries suggest spring or summer occupation based upon the recovery of a few "warm season" species such as migratory water fowl, and do not yield a high density of bison bone. We acknowledge that our small sample sizes might be introducing biases. However, since well-preserved bone has been recovered from this site, taphonomy does not explain the conspicuously low bison bone yield. Perhaps sites such as Bradshaw reflect a summer dispersal strategy whereby individual family units scattered to exploit the resources of the wetland-woodland microhabitats. These generalised foraging activities might also have permitted small-scale horticultural production. It is in these locales that we might expect to find direct evidence of horticultural production.

Research within the Lauder Sand Hills is preliminary. It is clear, however, that the high density of archaeological sites within these sandy wetlands indicates a focal point of Vickers Focus settlement in the northeastern Plains. Whether they were farmer-foragers or intensive foragers, it is clear that their settlement strategy was quite different from highly mobile hunting-foraging strategies used by such roughly contemporaneous societies as Blackduck. This is indicated by the sparse distribution of Vickers Focus sites throughout southern Manitoba. Most such sites are consistently found in isolated wetland zones associated with warm sandy soils. To our knowledge, no widespread distribution of Vickers Focus sites has been observed in North Dakota or Saskatchewan that might be indicative of seasonal mobility following bison migratory patterns. As forested well-watered locales similar to the Lauder and Rutledge Sandhills have not enjoyed much archaeological attention, there might be several other similar isolated localities containing dense concentrations of Late Plains Woodland sites.

We propose that circumscribed ecological conditions within the Oak Lake Aquifer region facilitated the perpetuation of an intensive centre-based settlement pattern reminiscent of eastern Woodlands societies. Vickers Focus settlement was focused upon specific microhabitats, providing a seasonally shifting resource base that sustained the population within a comparatively small area. In effect, Vickers Focus populations transplanted elements of their Woodland lifeway into a new area by seeking out suitable microhabitats. These forested wetlands offered abundant and reliable warm-season food supplies that included migratory waterfowl, aquatic plants and animals, and woodland resources. The forested wetlands offered predictable and seasonally clumped resources, and are reminiscent of favoured microhabitats throughout the Eastern Woodlands. The sandy, wellwatered sediments of the Lauder Sand Hills also could have favoured horticultural production. Such activities could have been conveniently embedded within regular foraging activities. While plausible, we repeat that our research has yet to yield evidence of horticulture.

The biotic productivity of wetland and forest habitats normally declines in winter. This might have forced intensive foragers to disperse widely in search of more diffuse food sources. An exception to this generalisation occurs when people are able to develop stored food surpluses sufficient to sustain the aggregated group through the season of scarcity. Throughout the eastern Woodlands, the development of stored food surpluses is generally associated with the development of horticulture, or economic intensification based upon rich and reliable wild food sources. The Lauder Sand Hills, located within the northeastern Plains, seem to have afforded new seasonal opportunities in the form of communal exploitation of bison that sought winter shelter and forage within the forested wetlands. By intensively exploiting seasonally clumped resources, a comparatively large population of Vickers Focus people appears to have sustained themselves within a small region over several centuries. We suspect that our site inventory within Makotchi-Ded Dontipi is overrepresented by winter bison procurement sites, and a consequent under representation of warm season occupations.

Movement of Plains Woodland and Plains Village populations westward into the eastern Plains is a well-documented phenomenon in the USA, but is rare in Canada. In the case of the more southerly regions, this migration involved exploitation of "linear microhabitats" represented by wooded and well-watered river valleys. Such microhabitats were used as a familiar base of operations for foraging, as a staging area for bison hunting, or as a sheltered base for horticultural production. In the more northerly Canadian Plains, forested microhabitats also existed in variety of situations that included river valleys, uplands, and also on land overlying aquifers. Interestingly, the only known concentration of Vickers Focus sites is within the sandy wetlands of the Oak Lake Aquifer region. This may be due to the hydrological conditions and locally extended growing season that allowed horticultural production. Alternatively, Vickers Focus people may have been strongly attracted to the area because of its seasonally balanced opportunities for warm season foraging and cold season bison hunting. In our view, such rich and diverse habitats are reminiscent of Plains Woodland homelands within the northern Mississippi River basin and the Prairie Peninsula.

This deliberate selection of wooded microhabitats within the northeastern Plains is consistent with the expansion of other groups out of the eastern Woodlands. Wood's (1971: 52-54) discussion of Cheyenne westward expansion serves as a useful analogy. He points out that in the late pre-contact and protohistoric periods, the Cheyenne may have been semi-sedentary farmer-foragers in Wisconsin and Minnesota. He proposes a time-transgressive shift over several centuries from this lifeway, to a Plains Village one involving semi-subterranean earth lodges, fortifications, corn horticulture and hunting. Wood cites Grinnell (1918, 1923) to suggest that these semisedentary villages were located along the lower Sheyenne River and adjacent portions of the Red River in southeastern North Dakota. By the mid 1700s, perhaps under military pressure from the Sioux, the Cheyenne moved farther west and adopted a typically mobile Plains equestrian bison hunting lifestyle. These transformations in settlement pattern and economy are characterised as gradual, and involved transitional shifts specifically involving microhabitats which are also ecologically transitional. The ethnohistoric documentation of Cheyenne cultural transformation suggests dramatic social and economic change over the course of little more than a century. It is probable that the direction and pace of these cultural adaptations were significantly influenced by indirect European contact, warfare and the development of equestrian Plains culture.

Given the comparatively coarse resolution of radiocarbon dating, it is difficult to archaeologically document similar transformations with Plains Woodland societies. That being said, we propose that similar transitions may characterise Vickers Focus adaptation to the northeastern Plains. There is a weakly expressed time-transgressive shift from east to west among the most northerly of Plains Woodland societies in the northeastern Plains. Assemblages with the strongest evidence of horticultural production are located in southeastern and south-central Manitoba. These deposits date slightly earlier than the more westerly sites that have not yielded evidence of horticulture, but which reveal a heavy reliance upon bison hunting. We propose that this pattern may indicate a general pattern of transformation of Plains Woodland societies as they moved westward and increasingly embraced a "Plains" style economy and settlement pattern focused upon bison hunting. Addressing the cultural historical relationship between Vickers Focus and Mortlach will be important in exploring these transitions.

Chapter 8

The New Face of the Clean Water Act

The United States is on the brink of a new era in water quality regulation. Newly-finalised total maximum daily load (TMDL) rules are bringing to life a long-dormant approach to the identification, prioritisation, and repair of the nation's polluted waters that promises to expand the gains in water quality secured by the Clean Water Act's first twenty-five years. Despite progress under the Clean Water Act (CWA), pollution or some form of habitat degradation continues to afflict thirty-six percent of surveyed river miles. Moreover, of the nation's impaired rivers and streams, less than ten percent are impaired primarily or secondarily by industrial point sources the Act's principal early target for pollution reductions along with municipal sewage treatment. These stark realities along with an abundance of litigation directed at the agency have prompted the U.S. Environmental Protection Agency (EPA) to explore the potential of the Clean Water Act's section 303(d) to promote a wide variety of new actions to protect the nation's waters.

A long-neglected part of the Clean Water Act (CWA), section 303(d) requires states to identify waters that are not in compliance with water quality standards, establish priorities, and implement improvements. Based in part upon the recommendations of a Federal Advisory Committee convened in 1996, the EPA issued proposed rules for implementation of the TMDL program in 1999. Amid substantial controversy the final rule was published in July, 2000.

The TMDL rule explains that "the TMDL specifies the amount of a particular pollutant that may be present in a waterbody, allocates allowable pollutant loads among sources, and provides the basis for attaining or maintaining water quality standards." Within this description lies a significant shift in the way water quality is regulated. Instead of the technology-based, end of pipe approach to point sources that has characterised CWA enforcement to date, the TMDL program promises an "ambient" approach to water monitoring and standards. That is, instead of a focus on releases from known sources of water pollution (i.e., monitoring discharges from discrete, identifiable pollution sources), regulation and reporting will increasingly be concerned with the in situ quality of waterbodies themselves. While this sounds like simple common sense, an ambient approach to water quality enforcement is largely untried. For a variety of mostly pragmatic reasons, federal and state programs have focused on the regulation of point sources via technology-based standards to secure effluent reductions. But the low-hanging fruit of low-cost, high-volume point source reductions has largely been harvested. Today, significant water quality improvement requires the expansion of controls to nonpoint sources of pollution.

While the rules have significant implications for point sources, the TMDL program's impact on nonpoint source regulation is its most important characteristic. Because ambient monitoring will find a large number of the nation's waterbodies to be impaired, and because nonpoint sources are a primary cause of that impairment, TMDLs will change the politics, economics, and implementation of water quality regulation.

This chapter provides a description and critical review of the EPA's TMDL rules. The analysis is not critical of the overall movement toward this type of regulatory approach. In fact, TMDL-based rules should be thought of as an inevitable step toward a mature phase of regulation in which all sources of water quality degradation are addressed. However, the review is critical in the sense that it takes a sober view of the significant challenges facing regulators. After all, there is a reason nonpoint sources have largely escaped regulation over the last twenty-five years. Federal authority to mandate nonpoint source controls remains weak. Implementation of the analytic tools required by the TMDL process will be costly and difficult. And conflicts are almost certain to arise due to the geographically interrelated nature of pollution sources and legal jurisdictions. Also, the new rules explicitly create incentives for pollutant trading across point sources and nonpoint sources. According to the new rules, "EPA is seeking to establish a market for pollutant trading, in the hopes of creating more effective and efficient mechanisms for restoring water quality." While laudable as a means to promote flexible, cost-effective discharge controls, water quality trading should be recognised as a tall order with numerous barriers to implementation.

This chapter does not address the debate over the CWA's net social benefits. Rather, it takes the CWA's legal requirements as given and considers the economic implications of movement toward a TMDL-driven regulatory system. Part II provides an overview of current water quality conditions, the history of the TMDL Program's development, and a brief description of the rules. Part III turns to a detailed description of completed TMDLs. Part IV explores a set of legal and economic challenges presented by the program's future implementation and makes recommendations for the ways in which the program can be made most effective. Part V considers the difficulties associated with water quality trading. The chapter concludes with a summary of the challenges facing the TMDL program.

The procedure by which water quality priorities are identified, the process that assigns responsibility for improvements, and the economic, political, and practical challenges of implementation, are all changing. The history of Clean Water Act implementation both its successes and failures helps to explain why the TMDL Program's time has come.

The Clean Water Act's Successes and Failures

Within ten years of the CWA's passage, water scientists and regulators reported significant improvements in water quality. An analysis of changes in water quality over the period 1974 to 1981 documented widespread reductions in lead and fecal bacteria concentrations, a reduction in industrial and municipal biological oxygen demand loads of seventy-one percent and forty-six percent, respectively, and some localised improvements in measures of dissolved oxygen deficit. To put the accomplishment in better perspective, these gains were achieved during a period when inflation-adjusted GNP increased twenty-five percent. The gains were largely due to CWA-related expenditures for the improvement of municipal sewage treatment. Between 1970 and 1985 the fraction of the U.S. population served by wastewater treatment jumped from forty-two percent to seventy-four percent.

In addition to improvements in municipal wastewater treatment, technology-based standards for industrial point sources began to yield large improvements. Starting in the early 1980s, point sources have had to meet effluent limits consistent with a variety of technology-based standards such as "best available technology economically achievable." Establishment of the National Pollution Discharge Elimination System (NPDES) permitting system has brought reporting, penalty, and anti-backsliding provisions to the regulatory scheme. Complementing this federal permit authority, section 505 of the CWA enables citizens to file suit for compliance with NPDES permits or EPA orders. Together, these enforcement tools have precipitated significant reductions in industrial point source releases.

For instance, between 1987 and 1990 toxic discharges to surface water fell from 417 to 197 million pounds per year, according to Toxics Release Inventory data. By setting limits on the discharge of more than sixty pollutants, including organic pollutants and heavy metals, the EPA Office of Water currently estimates that the program reduces conventional pollutant discharges by 108 million pounds and toxic discharges by twenty-four million pounds annually. A rough, but illustrative, benchmark of these collective improvements is offered by the EPA: "in 1972, most estimates were that only thirty to forty percent of assessed waters met water quality goals such as being safe for fishing and swimming. Today, state monitoring data indicate that between sixty to seventy percent of assessed waters meet state water quality goals."

There is at least an academic consensus that the CWA is one of the environmental movement's success stories, due to its success in motivating point sources controls. But to say that the CWA has improved water quality is not to say that further efforts are not indicated. In fact, the nation's waters remain significantly polluted.

Despite the CWA's achievements, a sobering implication of the above figures is that, while sixty to seventy percent of waters now meet state standards, thirty to forty percent still fall short. More specifically, according to the latest National Water Quality Inventory, thirty-six percent of the rivers and streams surveyed were partially or fully impaired, and water quality is "threatened" in an additional eight percent. Of the surveyed lakes, thirty-nine percent were partially or fully impaired, with water quality threatened in an additional ten percent. For surveyed estuaries, thirty-eight percent are reported to be partially or fully impaired, with water quality threatened in an additional four percent. Casting the data somewhat differently, the States have identified more than 20,000 individual river segments, lakes, and estuaries as polluted. This amounts to 300,000 miles of river and five million acres of lakes classified as polluted.

The data strongly suggest that improvements in water quality were disproportionately due to regulation of point sources. There is now widespread recognition that nonpoint sources are of principal concern. Consider the EPA's most recent ranking of sources contributing to water quality impairment; industrial sources, while present on this list, clearly do not predominate. Instead, a host of nonpoint sources, particularly from urban and agricultural runoff, loom large. Pesticide, fertilizer, and animal waste runoff from agriculture is the single largest contributor to the impairment of rivers and lakes. The private incentive of agriculture to ensure the largest yields via perhaps excessive application of pesticides and fertilizers is a classic example of an environmental externality, since much of the application inevitably migrates into common resource waterbodies.

Logging and construction activities, many of them on federal lands, are a significant source of sediment contamination, as runoff carries fine-grained soils from roads and construction sites into lakes and streams. In urban and suburban areas, watershed degradation is closely tied to increased population density and residential and commercial development.

In such areas the relatively impermeable nature of groundcover leads to rapid, unfiltered runoff from roadways and parking lots, chemically treated lawns, and commercial establishments. Increased attention is also being given to atmospheric deposition, where pollutants from airborne dust and industrial and commercial air emissions are absorbed by surface waters or precipitated via rainfall. In the Northwest, elevated water temperatures are having a negative impact on cold water salmon habitat. Finally, legacy pollution (i.e., pollution no longer being discharged) collected in sediments is a significant source of ongoing waterbody impairment.

One of the reasons these sources are such a significant problem is that they present serious implementation, monitoring, and enforcement challenges. But the water quality problems they cause can no longer be ignored, particularly given the significant reductions already secured from point sources. In this context, it is not surprising that political and legal pressure is being applied to the EPA, and in turn to the states, to make something of the regulatory potential contained in the CWA's TMDL provisions.

The Changing Politics of Water Quality

The seeds of this shift in regulatory emphasis have been in place since the Act's passage in 1972. Within section 303 lay provisions that called for an ambient water-quality driven (rather than end-of-pipe) approach to enforcement. The section calls upon the states to identify waters for which the point source controls elsewhere in the Act "are not stringent enough to implement any water quality standard applicable to such waters." States must priorities any waters so identified, based on analysis of use and severity of degradation, and establish TMDLs sufficient to bring the waters into compliance.

Section 303 provisions were largely ignored by states and the federal government during the first two decades of CWA enforcement. But nonattainment of water quality goals and the desire to bring more sources into the regulated sphere has led to a re-examination of the latent enforcement power contained in section 303. The importance of the section is, first, that it requires state-wide assessments and public documentation of water quality problems. As the public becomes aware of the "impaired" nature of the waters around their communities, this reporting alone will provide motivation for state regulatory efforts. Second, the TMDLs themselves appear, at least in principle, to imply that states must allocate pollutant load reductions to sources not currently covered by load restrictions. Since load reductions have been wrung from point sources over a period of twenty five years, and since the bulk of current impairment is due to nonpoint sources, any state seeking further load reductions at least on a cost-benefit basis will be led directly to nonpoint sources. In this way, the shift to ambient monitoring and standards almost necessarily leads to a greater emphasis on nonpoint sources.

Enforcement of section 303 clearly alters the politics of load reduction. The need to meet in situ water quality standards sets up a state-by-state confrontation between well-organised industrial interests which can claim to have already paid their pollution control dues and organised agricultural, silvicultural, and municipal interests that resist the "expansion" of CWA-driven requirements to their hard-to-solve nonpoint problems. An industrial and municipal constituency for nonpoint source controls can already be detected. According to the Director of the Association of Metropolitan Sewerage Agencies, "In the absence of nonpoint source controls, all these criteria drive tighter standards. All that's left is tighter limits on permits." This scenario is obviously of great concern to current point source permit holders. A minority (dissenting) contribution to the Federal Advisory Committee Report by a municipality representative and industrial source representative echoed the worry saying, "it is patently unfair to encourage states to impose further burdens on point sources merely because of the absence of federal enforcement authority over nonpoint sources."

The EPA's authority to implement the new TMDL rules is not accepted by all parties. Even if the new rules are successfully promulgated, many opponents have not accepted the changes quietly. Following publication of the proposed rule, members of a House subcommittee criticised the EPA for not requesting Congress to enact authorising legislation in support of the TMDL program. Some agricultural interests argue that the proposed TMDL rules "illegally link nonpoint source runoff to the federally dictated and enforceable TMDL program" by reading the CWA to cover only waters impaired by point sources. The EPA's position is that the CWA provides ample authority for the TMDL rules. In the Proposed Rule, the EPA directly addresses the argument that they have no authority to regulate nonpoint sources. "Section 319, a section that exclusively addresses nonpoint sources, provides clear evidence that Congress did not intend to limit the term `pollutant' to point sources." If a state fails to impose controls over an operation that is the source of an ongoing water impairment, the agency contends that the CWA provides it with authority to step in and issue Best Management Practices (BMPs) or other controls to reduce nonpoint pollution.

Litigation

The direct impetus for the new focus on section 303 was a series of court cases beginning in the late 1980s that challenged the EPA's oversight of the states' responsibilities. Until very recently, most states had failed to submit their assessments of polluted waters (if those assessments even existed at all), and had therefore failed to engage in the prioritisation of waters for cleanup, promulgation of TMDLs, and implementation of associated discharge controls called for under section 303. The history of this litigation has been usefully summarised by the EPA and others. As of late 1999, the EPA is under court order in seventeen states to establish TMDLS if the states fail to do so. In fourteen other states, claims have been filed (or notice has been given of intent to file) seeking to compel the EPA to establish

TMDLs. The Federal Advisory Committee and the new rules are a direct outgrowth of this litigation. The litigation has also had an immediate effect on the states' assessment and reporting of impaired waters. Idaho Sportsmen's Coalition v. Browner is illustrative. As of 1989, Idaho had submitted no list of "water quality limited segments" (WQLSs). As of 1992 the state had a list, but one with only thirty-six listed WQLSs. Using a "constructive submission" theory, the court found the EPA's approval of Idaho's compliance with section 303 to be arbitrary and contrary to law. Under court order, the EPA has since approved a list identifying 962 Idaho WQLSs. All states and territories have now at least submitted a list, with full EPA approval secured by thirty-eight states.

Brief Overview of the Enacted Rules

The revisions to the EPA's TMDL rules provide more specificity and structure to the operation of the program. The rules lay out a basic set of requirements for the states, including:

- Which waters are listed. While it is largely up to an individual state to determine which waters are impaired, relative to its own water quality standards, the rules require public review of the procedures by which a state generates its list. And EPA has final approval authority over listings and the methodologies used to derive them. According to the rules, the EPA's view is that "the section 303(d) list should serve as a comprehensive public accounting of all waterbodies impaired or threatened by pollution and pollutants, irrespective of the tool or mechanism being used to achieve standards."
- A format for listing and the assignment of priorities. The states are required to rank waters by their degree of impairment, assign waters on the list to categories that reflect their priority and progress toward meeting standards, and develop a "prioritised schedule" for attainment. Impaired waterbodies designated as public drinking water supplies and those whose water quality poses a threat to species listed under the Endangered Species Act, must be designated as high priority. Apart from that, the states are allowed discretion, subject to federal oversight. A range of factors may be considered in the development of the priority rankings. "These factors include immediate programmatic needs, vulnerability of particular waterbodies as aquatic habitats, recreational,

economic and aesthetic importance of particular waterbodies, degree of public interest and support, and State, Territorial, authorised Tribal, or national policies and priorities."

- Timelines. States are required to establish TMDLs for all waterbodies within ten years. TMDLs for waterbodies listed as "high priority" should be established first.
- The implementation of TMDLs themselves. All TMDLs must identify the pollutants contributing to impairment, and establish the load reductions, including a margin of safety, necessary to bring the water into compliance. The TMDL must also identify pollutant sources, including nonpoint sources. In addition, the TMDL must feature an implementation plan, with a list and timeline for actions, including monitoring and verification of compliance. The implementation plan also includes an "allocation" of load reductions to different sources. Local conditions ecological, political, and economic will be allowed to determine the nature of the allocation. According to the rule, TMDLs "provide for tradeoffs between alternative point and nonpoint source control options so that cost effectiveness, technical effectiveness, and the social and economic benefits of different allocations can be considered by decision-makers."

State-level TMDL development, particularly the allocation of load reductions, is likely to spur a host of innovative approaches to pollutant load reductions by drawing new sources and trading mechanisms into new control systems. But, to complement this many-flowers-will-bloom flexibility, the new rules codify a set of requirements meant to guarantee meaningful and comparable state initiatives. The rules are analysed in more detail below. While still subject to significant implementation questions, the rules should nevertheless be recognised as a significant new development in the nation's approach to water quality improvement.

Completed TMDLS

In response to the threat of litigation, or under court order, a number of states have moved forward with the development of loading standards for priority impaired water segments. We now describe Oregon's Columbia Slough TMDLs in order to concretely illustrate the components and practical implications of a TMDL analysis and plan.

The Columbia Slough TMDLs should not necessarily be viewed as representative of TMDLs generally, since the problems and solutions identified in TMDL planning are highly idiosyncratic to their location. The case has been selected because it involves a complex set of impairments and includes detailed analysis and concrete implementation plans by the state. In fact, the Columbia Slough TMDLs are notably specific in their description of the enforcement and implementation tools that will be employed to achieve water quality improvements.

Columbia Slough TMDLs

Once impairments are identified through the listing process, TMDL planning focuses on the identification of pollutant sources, the allocation of responsibility for source reductions, and implementation and enforcement of the allocations. These steps are illustrated in the following case description. Columbia Slough is a roughly nineteen mile long collection of mostly shallow water channels located on the Columbia River's floodplain near Portland. A variety of land uses industrial, residential, and agricultural occurs in the 40,000 acre area that drains into the Slough. The geographic scope of the TMDL was determined largely by hydrological characteristics such as drainage areas and tidal flows. The TMDLs for Columbia Slough were approved by the EPA in November, 1998.

The first step in the TMDL process is the listing of a waterbody as impaired. This process identifies the category of impairment (e.g., excessive algae or lead) and compares existing conditions to the relevant water quality standard. Impairment is determined according to criteria established and documented by the state. These criteria describe the standards for each category of impairment, the data required for analysis, and guidelines to ensure the quality of data analysis.

As an example, consider the state's framework for determining chlorophyll a impairment. The beneficial uses affected by impairment include water contact recreation, aesthetics, fishing, water supply, and livestock watering. The standards are expressed as numeric concentrations (e.g., mg/1). Water quality is considered limited, and thus subject to listing, if the 3 month average exceeds the numeric standard. Data used for the 1998 listing of waterbodies come from a variety of sources, including Oregon DEQ itself, as well as contractors and federal agencies. Standards and data vary by pollutant. For example, fish or shellfish consumption advisories issued by the Oregon State Health Division indicates impairment from toxic pollutants.

The Columbia Slough was placed on the Oregon Department of Environmental Quality's section 303(d) list due to a variety of impairments. The Slough was found to be in violation of standards for chlorophyll a, dissolved oxygen, pH, phosphorus, bacteria, and a range of toxic pollutants, including lead, dioxin, and PCBs. These pollutants individually threaten one or more beneficial uses of the waterbody including recreational fishing, boating, swimming, or support of aquatic life. Water column, fish tissue, and sediment data were used to identify the Slough as impaired. This data came from a variety of sources, including metropolitan agencies, the City of Portland Bureau of Environmental Services, Portland State University, and the U.S. Environmental Protection Agency (EPA). Historically, monitoring of the waterbody has been conducted only "sporadically," according to the DEQ.

The Columbia Slough case highlights the value of the listing process. As of 1998, Portland residents can refer to a readily available state analysis of a local waterbody that provides a range of recreational, aesthetic, and commercial values. That the waterbody suffers from ten distinct forms of impairment is an easily comprehended and presumably politically salient piece of information. Moreover, the listing process also promotes higher quality and more consistent data collection and analysis.

Their Contributions to Impairment Identified

With knowledge of impairment, states must design TMDLs that provide a defensible plan for source reductions to bring the waterbody into attainment. This requires knowledge of sources and the pathways by which pollutants are transported and deposited into the waterbody. Identification of sources and pathways requires a kind of holistic accounting exercise. To construct a source inventory, states will typically survey permitted sources (industrial and municipal point sources) and inventory land uses in the water body's drainage area to identify nonpoint sources, such as agricultural and urban runoff. In some cases, source identification requires the equivalent of detective work to connect ambient conditions with specific sources. For example, severe oxygen depletion was detected in the Slough following a large winter storm. This suggested that a source of impairment was aeroplane de-icing chemicals that ran into the Slough from Portland International Airport. In another case, elevated bacteria levels "substantially higher than those predicted via modelling" have directed attention toward the detection of illicit discharges.

In addition to source identification, the TMDL must estimate the effect of source reductions and other control activities on the Slough's water quality. In some cases, this process is straightforward. For example, an already permitted point source may discharge directly into the waterbody. If so, the source, transport and deposition mechanism and contribution to loading in the receiving waterbody are already known. In general, however, TMDLs will be geared toward waterbodies with multiple sources that migrate and enter the waterbody via a diverse set of pathways. Understanding the contribution of a particular source to loading in this more general set of cases is a significant technical challenge.

Sources of impairment in Columbia Slough include a complex mix of sewer overflows, urban runoff, landfill leachate, industrial discharges, sediments, and agricultural runoff. The complexity of making source determinations is compounded by a variety of interactions between ground and surface water, weather events, temperature, and water quality. " For this reason, source contributions are rarely known with certainty. Instead, the regulator must rely on models that attempt to capture the factors that affect the transport, deposition, and ultimate fate of pollutants in the waterbody.

The models used are determined by the type of data available and the nature of the system being described. For instance, stormwater discharges can be monitored for lead. This provides a rough gauge of the amount of lead deposition due to urban runoff. In turn, contributions to urban runoff by categories of sources can be estimated by modelling. Concretely, the land area in the drainage area occupied by industrial facilities or roadways can be used to estimate the fraction of lead runoff from those sources. The Columbia Slough TMDL includes a particularly detailed analysis of lead loading. Individual industrial, stormwater, and sewer point source permit holders have been identified and their lead discharges assessed. In addition, an analysis was performed to determine lead contributions by land use category. For each category, including residential, industrial, commercial, parks, and traffic corridor, the land area was calculated and the lead load (in pounds per year) was estimated. Consider the model used to calculate these loads for a single land use category: The contribution of lead from permitted industrial sites is calculated using the following equation: (Area x Annual Rainfall x Runoff Coefficient x Pollutant Concentration = Annual Pollutant Load). The annual rainfall for the watershed is given of the Portland MS4 permit application as 34.3 inches per year, and the runoff coefficient for industrial areas is given.

This example highlights the data requirements and assumptions that go into calculating a single number in a single analysis that is a component of a much larger evaluation of sources and relative pollutant contributions. The modelling techniques and data required for TMDL implementation pose a significant challenge. Accordingly, they also contribute significantly to the costs of implementation. Some simplicity and cost savings will undoubtedly be possible as states become more practiced in TMDL development and as more resources are devoted to the development of data and models for use in this kind of program. However, the degree to which data sources and modelling techniques can be standardised is limited. Each listed water segment is unique in its hydrology, transport pathways, pollutant sources, etc. TMDL development will invariably involve some site-specific analysis (even if the point of the analysis is simply to show that some set of standard data and techniques is applicable to the waterbody).

As an example of the site-specific nature of this analysis, consider the Columbia Slough TMDL approach to excessive nutrient loading. Because "point source loads are minor," the search for source reductions involves groundwater controls, such as the installation of sanitary sewers. However, because the effect of these reductions is extremely delayed and "uncertain," the plan calls for changes in channel and stream flow. The recommended changes are based on a site-specific analysis of natural flows, temperature, time of year, expected algae growth and flow management techniques available in the drainage area. The applicability of general modelling techniques to this kind of problem is limited. For Columbia Slough, the extent of documentation related to data, analysis, and methods is exemplary, as is the sophistication of the analysis itself. However, the volume and complexity of the modelling techniques imply an equally complex array of simplifying assumptions and potential shortcomings. In general, the TMDL process highlights data limitations and our ignorance of the physical, chemical, and ecological factors that determine the fate of pollutants.

Responsibility for Reductions

Having determined the contribution of source categories to impairment, the TMDL must assign load "allocations" (divisions of responsibility for reductions) that are expected to bring the waterbody into compliance. In the case of some, though not necessarily all sources, these allocations will require reductions in current discharges. Thus, the allocations determine responsibility for the technological or land management changes that will lead to improvements in water quality. For example, a currently discharging source, if allocated a zero wasteload, must totally eliminate its discharges.

In some cases, the allocation is assigned to specific sources, in others, to general categories such as a type of land use. Not surprisingly, allocations to specific sources tend be directed at point sources and general allocations directed toward to nonpoint sources. As in the determination of sources, the allocation process often involves a quantitative modelling exercise. Modelling is used to estimate the impact of a load reduction from a particular source on the receiving waterbody. Models may also be used to predict how changes in land use due to economic growth will add to the waste load.

Consider the allocations used to reduce Biochemical Oxygen Demand (BOD) in the Slough. Recall that the two primary sources contributing to the oxygen deficit are Portland International Airport and urban runoff. Oregon allocated responsibility for reductions by considering the relative baseline contributions to the oxygen deficit. These relative contributions were determined, via modelling, to be 3.8 lbs. of BOD airport load for every pound of urban runoff load. This ratio provided the basis for the allocation of reductions. Within these broad categories further divisions were made. For de-icing, eighty-nine percent was assigned to the airport and the remaining eleven percent to the Oregon National Guard. For urban runoff, forty-six percent was assigned to permitted stormwater dischargers and the remainder to future growth and non-permitted sources. In the course of the public comment process the allocation was explained as follows: "DEQ feels that allocations based on the relative contributions of the pollutant are appropriate for a situation in which one source causes most of the water quality impairment." Note that the rationale for the allocation makes no reference to the costs of discharge controls. Instead, equity seems the underlying motivation. No explicit rationale need be, or is, given.

Allocations are also motivated by pragmatism. In several cases, the permitted allocation is equal to the baseline contribution of a source, implying no need for source reductions. These examples include upstream sources of bacteria (which, because they are upstream are not controllable via this particular TMDL) and atmospheric deposition of lead (uncontrollable for similar reasons). Interestingly, specific point sources are in some cases also exempt from load reductions. This is true presumably because point sources are already subject to "best available" controls on releases. For example, a landfill generating lead-contaminated leachate was not required to achieve reductions in excess of its baseline allocation.

Discharge reductions in Columbia Slough are to come from a variety of sources. For bacteria, reductions are achieved by eliminating combined sewer overflow (CSO) discharges and eliminating illegal raw sewage sources. Both received a load allocation of zero. The primary reductions in lead loadings are allocated to sources of urban runoff. These reductions are to come from changes in industrial stormwater permits and implementation of BMPs applied to commercial, industrial, and traffic corridor land uses. For nutrients, reductions from a single agricultural point source and drainage district water flow management changes are required. Direct reduction in organic contaminants, such as DDT and dioxin are not possible since the release of these pollutants is now prohibited. Instead, the TMDL focuses on erosion in the drainage area. Stormwater transport of contaminated sediment is the primary transport and deposition mechanism. The stormwater allocation is reduced. Moreover, of the total organic allocation, thirty-three percent is devoted to future growth, since construction promotes erosion and is expected to increase loads. Reductions are to be achieved by implementation of erosion control BMPs.

How Will Implementation and Enforcement of the TMDL Be Achieved? With the allocation of loads, and the reductions in discharges they imply, TMDLs must then specify the policies that will be used to achieve those allocations. How are allocations to be monitored? What enforcement tools can be brought to bear to ensure compliance? Is private or public funding available for the technological and management changes implied by the allocations? The more specific and helpful a TMDL is in response to these questions, the more likely are its load reductions to be translated from aspirations into reality.

The Columbia Slough TMDL is noteworthy in that it provides particularly detailed descriptions of implementation activities. Implementation takes a variety of forms, including monitoring, revisions to NPDES permits, public capital projects, and the use of BMPs. Monitoring requirements include an ongoing study of airport-related BOD, a survey of area septic systems to identify sources of bacteria, and monitoring of toxic runoff associated with environmental cleanup sites. Revision of the airport's NPDES permit is identified as a way to promote more effective treatment of de-icing runoff. Public capital projects that will lead to reductions include the construction of sanitary sewers, removal of cesspools, construction of treatment facilities for CSOs, and separation of sewers from stormwater systems. Both private landowners and local government entities are required to use a variety of BMPs, including redesign of roof drains to avoid bacteria discharges from septic systems, techniques to control erosion and the transport of contaminated sediment, and changes in flow management by the area drainage district. Renewal of stormwater permits will also be contingent on evidence of BMP compliance.

The plans are also explicit in their assignment of responsibility for monitoring activities and BMP implementation among government entities. Memorandums of agreement between designated management agencies form the basis for this division of authority. Responsibility is divided between the DEQ, a collection of municipal governments, county government, and the state Departments of Agriculture and Transportation. In addition, DEQ lists a set of possible funding sources for TMDL related attainment projects, available to both private and public entities. On paper, the Columbia Slough TMDL illustrates the promise of an ambient water quality-driven regulatory approach. The TMDL identifies impairments and their sources, allocates responsibility for source reductions and management improvements, and offers a detailed plan for monitoring and implementing improvements. This marks an ambitious expansion of efforts relative to conventional water quality regulation. The TMDL expands waterbody and source monitoring and the techniques used to analyse pollutant transport and deposition. The plan also imposes regulatory requirements on previously unregulated sources and mandates changes in land and water management practices throughout the drainage area. Finally, the TMDL notifies the source of those problems and identifies the private and public parties responsible for improvements.

Other Examples

One of the distinguishing features of Columbia Slough is that agricultural nonpoint sources are not a significant source of impairment. As previously noted, agricultural nonpoint sources will be featured in many TMDLs, due to the prevalence of excessive nutrient loadings in many of the nation's waterways. A brief description of nutrient TMDLs is therefore instructive. Like the Columbia Slough analyses, TMDLs for nutrients involve modelling exercises to determine source contributions from agriculture, urban nonpoint sources, and point sources all of which can contribute to nitrogen and phosphorus loads.

North Carolina's plan for nitrogen reduction in the Neuse River basin is illustrative. The TMDL estimates the relative contributions of point and nonpoint sources and calls for thirty percent reductions in nitrogen loads from both point and nonpoint sources. For agricultural nonpoint sources, the TMDL proposes to achieve reductions through farmer participation in local planning exercises or compliance with BMPs (specifically, installation of forested riparian areas or vegetative filter strips). Other initiatives included in the TMDL are a requirement that those applying fertilizer to areas of more than fifty acres receive training in nutrient management.

A characteristic of several nutrient TMDLs is that reductions are sought from both point sources and from nonpoint sources. Maryland requires NPDES permitting changes for sewage treatment in the Port Tobacco watershed. California calls for revision of NPDES permits to meet nutrient reductions in San Diego Creek and Newport Bay. Delaware's Nanticoke River TMDL requires nutrient removal at three wastewater treatment plants, and Delaware's Indian River TMDL calls for the total elimination of point source discharges. Interestingly, several TMDLs are also counting on reductions in atmospheric deposition due to more stringent CAA controls on emissions of nitrogen oxides in order to meet nitrogen load reduction goals.

In all of these cases, however, the net is also cast toward currently unregulated sources, including agricultural nonpoint sources. All of the aforementioned TMDLs set nonpoint allocations. The San Diego Creek TMDL proposes to extend discharge requirements to small, unregulated nurseries. It also requires the development of nutrient management plans for all agricultural operations not regulated by waste discharge requirements. The Maryland Port Tobacco TMDL's nonpoint allocation assumes no additional requirements beyond existing BMP practices. It does suggest that existing state programs will support the implementation of future nonpoint source controls, however. Delaware's Nanticoke River and Indian River TMDLs require plans to institute agricultural BMPs in order to achieve those TMDLs' nonpoint allocations. And as noted above, North Carolina's Neuse River TMDL explicitly requires expanded use of agricultural BMPs. The problems associated with translating nonpoint allocations into concrete improvements are explored below. Completed TMDLs do indicate that a set of previously unregulated, unmanaged sources is experiencing, at the very least, heightened regulatory scrutiny.

Long-Run Implementation of the Programme

Having described the scope and promise of the new TMDL rules, we now turn to issues raised by the long-run implementation of the program. The federal government and states have broad, if somewhat untested, legal authority for expansion of controls to previously unregulated sources. After a description of their respective authority we turn to an exploration of policy tools available to promote source reductions and beneficial land and water management practices. Then, we describe a set of scientific, legal, and practical challenges that will be faced by states as they seek approval for and implementation of their TMDL programs.

Federal Authority and Nonpoint Sources

The Clean Water Act prohibits discharge of a pollutant without a permit. This broad prohibition is subject to a very important limitation, however. It applies in a legal sense only to discharges from point sources. One way to expand federal legal authority is the semantic strategy of simply relabelling nonpoint sources as point sources. In fact, the EPA proposed (but withdrew) an expansion of the definition of point sources to large-scale feeding operations and certain aquaculture and silviculture practices as part of its 1999 proposals. In some cases, even smaller agricultural operations may be reclassified as sources subject to NPDES permitting, or threatened with such reclassification as an incentive to improve management practices. But despite efforts to expand the universe of sources subject to the CWA's permitting, there remains a large universe of sources that resist simple reclassification.

As a result, the CWA's powerful, permit-driven mode of regulation does not apply to the primary cause of current water degradation.

Some federal leverage exists, but it is more indirect than that provided for by the CWA's point source provisions. For example, section 319 of the Act addresses waters impaired by nonpoint sources. The section requires states to submit reports of such impairment, identify broad categories of sources, and identify BMPs and other measures to control "to the maximum extent practicable" pollution from those sources. States are left with the discretion to determine whether control efforts are to be regulatory or nonregulatory, compelled or voluntary. Nowhere in the section is an explicit enforcement authority granted to the federal government. In principle, the federal government could withhold a state's NPDES permitting authority if the state fails to follow through on section 319's requirements. But this extreme measure is not a substitute for a more explicit, practical federal ability to compel nonpoint reductions.

More subtle forms of federal influence are available. Appropriations are available to states under section 319 and section 208. The threat of funding withdrawals, while a relatively weak incentive given the limited nature of funds at stake, is available to the EPA. Section 208 instructs states, among other things, to develop a process to identify agricultural and forestry nonpoint sources. These sources include return flows from irrigated agriculture, runoff from manure disposal areas, and land used for livestock and crop production. The section sets forth procedures and methods (including land use requirements) to control such sources to the extent feasible and provides revocable funding for the process. EPA efforts to motivate state nonpoint initiatives are illustrated by an EPA memo to regional administrators, calling on them to "focus substantial grant dollars toward those states that are providing reasonable assurances that nonpoint source load allocations established in TMDLs will in fact be achieved." The memo also suggests that the administrators may "deny or revoke a state's enhanced benefits status under the new section 319 nonpoint source guidance." These types of funding incentives indicate the EPA's concern with state nonpoint program development. They also signal the lack of more substantive authority on the part of the EPA to motivate that development.

Federal authority over the management of federal lands can be used to promote nonpoint source improvements. The Forest Service and the Bureau of Land Management, in principle, can use their authority to ensure adequate water monitoring and the implementation of BMPs for grazing, logging, and road construction activities. In fact, both agencies have mandates to manage public lands for multiple uses, including recreation, fish and wildlife, and watershed uses. Also, section 401 of the CWA empowers the States to review facilities or activities that require a federal permit. Thus, states can compel federal agencies to manage lands in a manner compatible with water quality maintenance and improvement. Finally, the Department of Agriculture's Conservation Reserve and Wetland Reserve Programs in principle could be harnessed to promote BMPs on agricultural land. Access to agricultural price support and land retirement programs could be made conditional on the adoption of pesticide and fertilizer reductions, though this is not current practice.

Regarding the TMDL program itself, it is clear that primary authority for policies to implement TMDLs resides with the states. States have primary responsibility for developing their own lists of impaired waters and they are granted wide latitude to determine their own priorities and implementation plans. However, recent litigation and the EPA's own posture suggest that there must, and will, be a strong federal oversight of these state programs. The proposed rules outline the agency's vision of its own authority. According to the rules, "EPA has strong and diverse authorities to implement controls over nonpoint sources in the event that EPA were to disapprove a TMDL submitted by a State and to develop a TMDL for the impaired water." This assertion sounds somewhat defensive, and should, since as previously argued federal authority to compel nonpoint controls is limited. However, the rule identifies section 504 as one particular source of federal authority. Under section 504, the administrator can compel action when there is an "imminent or substantial endangerment to health or welfare of persons." As an example of such an endangerment, the rule suggests a community's inability to market contaminated shellfish. But in general, the federal TMDL program is not self-implementing. As the proposed rule acknowledges, load allocations for nonpoint sources are not directly enforceable under the CWA. "With respect to nonpoint sources the load allocations in a TMDL are only `enforceable' to the extend they are made so by state laws and regulations." Despite the federal government's clear role in prompting and overseeing state action under section 303, state law will ultimately determine the effectiveness of the TMDL program's long-run implementation.

State Authority and Nonpoint Sources

The litmus test of the TMDL program's success will be its ability to promote more effective nonpoint controls at the state level. Nonpoint sources will increasingly occupy the attention of States as they identify impairments and sources and develop plans for water quality improvement. But as argued above, nonpoint sources historically have not been the prime target of regulation. Moreover, there is likely to be political pressure at the state level to avoid an expansion of controls to this category of sources. To date, state initiatives have relied heavily on voluntary, unenforceable measures, particularly with regard to agricultural runoff. Taken together, these factors raise concern regarding the states' willingness to meaningfully compel reductions in nonpoint loads. The EPA's proposed TMDL rule anticipates this concern by requiring "reasonable assurance" of implementation. These reasonable assurances require states to specify implementation policies, the timing of controls or incentives, analysis of the likely effectiveness of policies, and funding sources.

For point sources, consistency with NPDES permits is considered a valid "assurance" of implementation. For nonpoint sources, the most direct assurance of implementation is the availability of state laws to compel nonpoint controls, and the willingness to enforce those laws. There are in fact a large number of state statutory provisions that could be called into service. These laws include general prohibitions against pollution discharges, enforcement actions triggered by fish kills or threats to public health, sedimentation and erosion laws, and laws designed to protect specific areas for conservation. Most states have access to such statutory levers, which can be directly cited as a form of reasonable assurance. In addition, regulations governing stormwater runoff, zoning, and other land use ordinances would likely qualify. Assurance could also be demonstrated by management contracts between government agencies and land users, memoranda of understanding between government entities, and bonding requirements to ensure appropriate land management practices. A variety of approaches will arise as state regulators seek innovative ways to motivate quality improvements. As an example, there are proposals in Washington State to limit irrigation water allocations to farmers who fail to take precautions against sediment runoff. In Iowa, a pilot crop yield insurance plan reduces the risk to farmers of under-performing crop yields when they decrease fertilizer applications.

These examples are indicative of the kind of policy innovations that can be harnessed by states to compel or otherwise promote nonpoint load reductions.

On one hand, the long list of existing state enforcement tools is reassuring. On the other hand, that long list has thus far failed to yield adequate water quality in many of the nation's streams, lakes, and estuaries. In some cases, the failure is due to explicit exemptions for certain nonpoint sources. In other cases, the law may simply be unenforced. Many of the states with general discharge provisions explicitly exempt agriculture. Massachusetts prohibits the discharge of any pollutant without a permit, but agricultural nonpoint discharges are exempted. Exemptions are also found in more targeted regulations. In Ohio, land development that disturbs the soil cannot occur without state approval. But, this requirement does not extend to agricultural operations. The TMDL-driven need for load reductions may erode some of these statutory exemptions as point sources permitees exert political pressure of their own to avoid further, costly point source reductions, and environmental interests focus greater attention on impairments. Dormant provisions may also be called into more active service by state program administrators who must ultimately demonstrate real pollution reductions under section 303.

Best Management Practices as Enforceable Standards

Best Management Practices are a central feature of most nonpoint source control programs. BMPs are management standards that guide forest, agriculture, construction, and other activities in order to reduce nonpoint runoff. BMPs are based on the practical experience of land managers and improvements in the scientific and technical understanding of the relationship between land management practices and environmental impacts. In agriculture, examples include the installation of buffer strips along stream beds, adequate fencing to keep livestock from directly soiling surface water, placing sheds over manure piles to minimise runoff, and low chemical intensity pest control techniques. Forestry BMPs include harvest and road construction planning to avoid soil erosion from trails, roads, and stream crossings. Construction BMPs have a similar focus on techniques to minimise sedimentation due to erosion caused by soil disturbance. BMPs for municipalities include procedures to minimise the impact of road salting on urban runoff. Depending on their application, BMPs can represent informal rules of thumb or be subject to approval by a government entity, such as a conservation district, state, or the U.S. Forest Service. As an illustration, Florida employs BMPs approved by a variety of organisations, including the U.S. Department of Agriculture Natural Resources Conservation Service and the Florida Department of Agriculture and Consumer Services (FDAC). FDAC has issued specific conservation practices for the purpose of protecting Florida's water resources (including streams, lakes and wetlands) from pollution associated with forestry operations. These BMPs were originally designed in the mid-1970s in response to the CWA, but were revised in 1993 with the assistance of representatives from state and federal government, universities, the forest industry, and environmental groups. The recommended practices are detailed and depend on the size and type of waterbody involved, the local soil type, and the general potential of the site for erosion and sedimentation. A BMP Technical Advisory Committee meets biennially to evaluate the status and progress of BMP implementation and effectiveness.

Section 208 of the CWA, which mandates state water quality plans, calls on states to describe in its plans the BMPS "which the [state] agency has selected as the means to control nonpoint source pollution where necessary to protect or achieve approved water uses." The section outlines the desired characteristics of BMPs for a variety of nonpoint source categories. BMPs represent the nonpoint analogue to end-of-pipe controls on point sources. They identify the technologies and techniques that lead to pollutant reductions. In some cases, states use BMPs as an aspirational goal or background threat, rather than as an enforceable standard. Colorado law calls for the development and approval of BMPs by the state department of agriculture, but defines BMPs as "any voluntary activity, procedure, or practice." Thus, BMPs are not directly enforceable. Colorado does encourage their use to avoid the possibility of future regulation if "continued monitoring reveals that rules and regulations are not preventing or mitigating the presence of the subject agricultural chemical to the extent necessary." But, reliance on voluntary efforts should be viewed with skepticism.

In other cases, BMPs are directly enforceable. Oregon law requires the State Forestry Board to develop and require the use of BMPs "as necessary to insure that to the maximum extent practicable nonpoint source discharges of pollutants resulting from forest operations on forestlands do not impair the achievement and maintenance of water quality standards for the waters of the state," and criminal and civil penalties can be levied for failure to use these BMPs. This regulatory "stick" is complemented by a significant "carrot," however. Forest operations conducted in accordance with BMPs can use such management as a defence against claims of water quality standard violations.

In Maryland, agricultural operations must employ BMPs under soil conservation district plans to protect nontidal wetlands. Similar requirements are imposed on agriculture in the Chesapeake Bay watershed. Federal law can compel the implementation of BMPs, particularly on federal lands. In addition, states can independently force the federal government to make its financial assistance programs, permits, licenses, and development projects conform with state nonpoint control programs, which may feature BMPs.

Recent cases illustrate a growing judicial awareness of the importance of BMPs. In Sierra Club v. Martin, the U.S. Forest Service's adherence to BMPs was at issue. In Blue Mountains Biodiversity Project v. Blackwood, BMPs associated with a site assessment were found to be inadequate given the characteristics of the site. Idaho Sporting Congress v. Jemmet, tested the compliance of a National Forest timber sale contract with applicable BMPs. These cases suggest that BMPs increasingly can, and will, be examined to determine compliance with statutory land management requirements.

The Technical Basis for Listing Criteria and Load Modelling

The TMDL program requires states to develop scientifically and legally defensible data collection procedures, listing criteria, and watershed modelling tools. TMDLs may be challenged if the implementing state agency fails to adequately define and employ these technical duties.

In terms of listing waters as impaired, the proposed rule requires states to include with their section 303(d) lists a description of the methodology and factors used to priorities and list waters as impaired. As an example, the methodology description can explain how the number and severity of "exceedences" of numeric chemical criteria translate into an impairment. Similar requirements are applied to TMDL load reduction plans. Implementation plans are required to contain monitoring and modelling procedures that will be used to gauge the effectiveness of load reduction actions. States are also required to explain their approach to "assessing the effectiveness of best management practices and control actions for nonpoint sources." These requirements may slow the process of TMDL implementation since they require the adoption of what may, for some states, be new techniques.

Pollutant sources, dissatisfied with their designation, may also seek relief from TMDL load reductions by challenging a state's modelling tools, water quality criteria, and data collection procedures.

Recent litigation has focused on one state agency's inability to promulgate lawful section 303(d) listing criteria and TMDLs associated with nutrient water quality standards. In the case, the South Carolina Department of Health and Environmental Control's (DHEC) listing standards and associated TMDLs were voided by an administrative law judge.

The agency was found to have violated proper public notice and other procedural safeguards designed to subject its standards and models to technical and legal scrutiny. According to the court, "DHEC has pursued its mission with unpromulgated regulations that should have been, but were not, subject to the scrutiny of DHEC's Board, the South Carolina General Assembly, and the public."

While federal water quality standards are available, they are published by the EPA only as non-regulatory guidance. The promulgation of water quality standards lies squarely within the jurisdiction of state law. Moreover, the CWA requires that criteria adopted by states must be consistent with their own laws governing how regulations become law.

DHEC's failure was in inadequately translating narrative state water quality criteria into numeric standards for phosphorus, inorganic nitrogen, and chlorophyll a. Narrative standards are common in state statutes and, as the term suggests, are general, verbal descriptions of required water quality. In South Carolina, one such narrative standard reads that waters shall be free from "waste in concentrations or combinations which interfere with classified water uses, existing water uses, or which are harmful to human, animal, plant, or aquatic life." The court explained that the state must "for both practical and regulatory reasons, 'translate' its narrative water quality criteria into numerical values when making section 303(d) listing determinations." The court's ruling found that the state failed to make this translation in a way that satisfied procedural requirements. The standards being used by DHEC failed to have "clear bounds and a rational basis for their implementation."

The agency cannot relist the waters under section 303(d) until the U.S. EPA approves the "translation procedure" for converting South Carolina's narrative criteria into numeric criteria. Perhaps more significantly, the judge's ruling also requires the agency to conduct a case-specific assessment of the waterbodies and sources in question. Assessment of the waterbodies is essentially a data collection and evaluation exercise.

The analysis of sources is typically more complicated. Specifically, the agency must conduct an evaluation "of the point and nonpoint source nutrient loadings, of other possible causes or contributors to water quality impairment, and of whether Section 303(d) listing can be avoided by DHEC's full implementation of existing point and nonpoint source controls, including full implementation of BMPs at nonpoint sources." This kind of exercise is fraught with technical difficulties.

Analysis of loadings and the effect of load reductions requires a watershed model that captures transport process (such as infiltration and runoff), ground water and surface water interactions, pollutant accumulation and decay, and instream mixing. In the case of nonpoint source loads, the science is relatively undeveloped due to the complexity of interacting systems involved. Knowledge of the relationship between control practices and loadings is particularly poor. According to an EPA supporting document for TMDL development, a key challenge facing agencies is "the lack of highly developed, scientifically sound approaches to identify problems in watersheds and to predict the results of potential control actions on water quality. While a wide variety of models are available, each come with limitations on its use, applicability, and predictive capabilities."

Consider one particular, and relatively narrow technical issue: the interaction between ground water and surface water quality. The "flows" of groundwater into surface water (or vice versa) are themselves highly uncertain and may occur over a period of decades. This makes cause and effect determinations nearly impossible. The long time lag also limits the ability of researchers to measure the effect of control actions on receiving water quality. Moreover, the conditions under which surface and ground water interact have a crucial effect on water chemistry (acidity, oxygen content, temperature) and biological conditions that ultimately affect water quality. Unfortunately, according to the U.S. Geological Survey, "research on the interface between groundwater and surface water has increased in recent years but only a few stream environments have been studied, and the transferability of the research results is limited and uncertain."

A lack of scientific certainty will in and of itself not legally hobble TMDL plans, since certainty is not a prerequisite for program implementation. Uncertainty does place a premium, however, on administrative procedures that provide the greatest possible level of scientific credibility to standards, models, and data collection. Accordingly, the technical details of state TMDL programs will need to engage in ongoing notice and comment procedures and evaluation by expert panels. This is likely to be a source of both significant upfront and long-run program costs.

Jurisdictional Conflicts

Section 303 provides a great deal of state discretion to determine standards and implementation strategies. This latitude, together with the lack of correspondence between state boundaries and watershed boundaries, raises the possibility of jurisdictional conflict over TMDLs. Most obviously, downstream water segments may inherit water quality problems from upstream sources in other states. Standards differ across states, sometimes to a significant degree. Less stringent water quality standards and less effective implementation upstream can alone create impairments in a downstream state with stricter quality standards.

For this reason, the proposed rules require states to identify a process for resolving disagreements with other jurisdictions. The rules also state that the EPA may establish TMDLs "when interstate or international issues and coordination needs require EPA to assume a leadership role." This will be particularly true in the case of large rivers or boundary waters. The agency also sees a role for itself in determining "equitable upstream / downstream allocations that account for loadings to downstream waterbodies like the Chesapeake Bay from far away upstream sources."

The need for a federal coordinating presence is likely to be particularly acute when conflicts arise due to atmospheric deposition. Atmospheric deposition occurs when airborne pollution is deposited directly onto the surface of the waterbody, or indirectly onto land and waters within the watershed. Because atmospheric deposition is a major source of water impairment, section 303(d) inextricably links state compliance with controls mandated by the Clean Air Act. According to the EPA, eighty percent of the Delaware Bay's mercury load, forty-six percent of Tampa Bay's cadmium load, and twenty-seven percent of the Chesapeake Bay's nitrogen load are due to airborne sources.

The significance of these atmospheric loadings is that they implicate sources over a huge geographic area. Deposition in the Chesapeake Bay is a particularly pertinent example. Studies based on data from the National Acid Deposition Assessment Program suggest that only twenty-five percent of air deposition in the Chesapeake watershed originates from sources within the watershed (which contains areas in six states). The Chesapeake NOx "airshed" which defines the geographic range of airborne nitrogen sources to the Bay, covers areas in thirteen states plus the District of Columbia. Given the significance of air deposition in that watershed, and the broad geographic range of sources, there is the distinct possibility of jurisdictional conflict that necessitates federal intervention.

TMDLs and Water Quantity Law

As they move toward implementation, TMDL rules will increasingly highlight the artificial distinction between water quality and quantity issues, particularly in the West. Water quantity decisions, which are controlled primarily by state law, often have a direct impact on water quality. For instance, changes in stream flow affect the transport of pollutants through a waterbody. Also, the amount of water taken or returned to a waterbody may significantly affect the dilution of pollutants in that system.

Finally, water supply often determines the suitability of a waterbody as habitat for fish or other species. In fact, reduced stream flows can constitute "water pollution" under the CWA. At a practical level, TMDLs will often have to account for seasonal changes in flow in order to set appropriate loadings consistent with states' water quality standards. Because of these interactions, water quantity decisions relating to irrigation, damming, reservoir management, basin-to-basin trades, etc may result in changes in the waters' section 303(d) status. Correspondingly, TMDLs will in some cases constrain water transfers involving impaired waterbodies.

As argued above, a wide variety of unresolved legal issues impact the implementation of the TMDL program. The interrelationship between water quantity decisions and water quality conditions will present particularly challenging issues. Water quantity law has a well-deserved reputation for complexity.

Water rights have both public and private characteristics, water law is largely state-determined, different laws govern groundwater and surface water (even though they are physically interdependent), and entirely different systems for establishing rights are used in different regions of the country.

In terms of interjurisdictional conflict, states have a long history of conflict over water quantity apportionment. This type of conflict will likely increase under the TMDL program as water quality impairments become binding enforcement problems. One consequence of quality concerns may be an increasing reliance on water quantity acquisitions to preserve stream flows, an approach endorsed by the TMDL federal advisory committee. Federal acquisitions have already occurred due to concerns over species habitat. Many states already have in place the legal foundations necessary to allow purchase and trade of water rights to preserve instream flow.

The Location of Sources

TMDLs will raise the costs of, and in extreme cases create a prohibition against, source development in watersheds suffering from impairment. New source restrictions, or any other policy creating barriers to new sources, are a necessary element of any policy geared toward water quality improvements. However, an unintended consequence of TMDLs is that they may encourage the migration of point and nonpoint sources to areas that are not currently associated with water impairment. In fact, the more successful policies are in limiting loadings in impaired areas, the greater will be the incentive of dischargers to relocate to areas where TMDL driven restrictions are not as binding or where control costs are not as high.

Economic Considerations And Pollutant Trading

A distinctive feature of the EPA's approach to TMDL rules, and its watershed policy more generally, is an emphasis on trading to achieve "common-sense, cost-effective solutions for water quality problems." In principle, trading is highly desirable. Trading allows sources with responsibility for discharge reductions the flexibility to determine where those reductions will occur. The financial incentives built into a trading scheme lead naturally to a situation in which the costs of pollution control are minimised.

Pollution permit markets, in a decentralised manner, assign control activities to the parties whose control costs are least. This is economically efficient. It is also politically attractive, since it minimises compliance costs for a given pollutant reduction goal. In terms of water quality regulation, trading's desirability arises due to a vast disparity in pollutant control costs across sources. This is particularly true when nonpoint sources are considered. Nonpoint source BMPs are thought to be a particularly cost-effective means to achieve water quality improvements.

For example, if point sources with high control costs can purchase cheaper controls from nonpoint sources, the control cost savings may be significant. A wide variety of trading possibilities is contemplated by the EPA. A justification for the proposed new source offset rule was that "EPA believes this proposed requirement will serve as a catalyst for the establishment of a trading market between large new dischargers and existing dischargers undergoing a significant expansion, and existing point source dischargers or nonpoint sources." Point-point, point-nonpoint, and nonpoint-nonpoint trades are all envisioned.

The cost savings and flexibility provided by trading are the principal motivations. Potential savings from point-point trading alone are estimated to be as high as $1.9 billion per year. The theoretical desirability of pollution trading is accompanied by a host of sobering practical challenges, however, and the history of actual point-point trading to date under the CWA is very limited. More limited still is persuasive evidence that trading implementation can result in significant cost savings.

Trading under the Clean Water Act

The history of water quality trading can be summarised relatively briefly. A recent summary of trading and offset programs lists thirty-seven such programs at various stages of development, the majority of which (twenty-six) have not actually resulted in trades. Notable trading experiments include the now-defunct Fox River BOD trading program in Wisconsin. Preliminary estimates of the savings from this point-point trading program were $7 million a year for the participating firms. Unfortunately, and in stark contrast to the prediction of cost savings, the program only produced a single trade.

In Colorado, the Dillon Reservoir trading system allows trade between point sources and nonpoint sources to reduce phosphorus loadings in the reservoir. Point-nonpoint trading activity has not actually materialised. However, two trades between nonpoint sources have occurred as nonpoint sources have become the primary remaining source of discharges.

The Minnesota Pollution Control Agency has approved a single trade which substitutes nonpoint phosphorus source reductions for a point source reduction on the Minnesota River. In North Carolina, the Tar-Pamlico plan calls for point source nutrient reductions. Point sources must meet a cap on discharges or make a mandatory financial contribution to a nonpoint source reduction fund used to implement agricultural BMPs. To date, the point sources have met the annually decreasing cap, primarily through improvements to treatment facilities. These and other trading programs indicate state and EPA willingness to experiment with the trading approach. However, they are not particularly inspiring examples of trading are ability to achieve significant cost reductions, or enthusiasm from the regulated community. What is remarkable is how little actual trading has arisen from these programs.

The Barriers to Water Quality Trading

An effective trading system requires several fundamental ingredients. Among them are: first, a sound means of enforcement to ensure that commitments (load reductions) are adhered to; second, a legal foundation that allows control flexibility sufficient to generate financial gains to participation; and third, an administratively straightforward process for participating in the market. In the case of water quality trading, many of these ingredients currently are missing.

Perhaps the greatest barrier to water quality trading is the sheer complexity of factors that determine watershed loadings. Even if a source's releases are perfectly known a tall order in itself, particularly for nonpoint sources the transport and deposition of releases are subject to numerous uncertainties relating to geography, hydrography, and weather conditions. Once present in a waterbody, a pollutant's contribution to impairment is often a function of the waterbody's assimilative capacity (which is itself a function of rainfall), temperature, salinity, acidity, and other localised chemical characteristics. This produces several problems for a trading program because it makes it difficult to draw causal relationships between specific sources and water quality problems.

This causation problem creates an obvious monitoring problem, unless releases can be monitored at each potential source. It also makes it nearly impossible to quantitatively relate control practices (e.g., a reduction in fertilizer application) to loadings. Establishing those relationships is necessary to establish appropriate trading ratios. Trading ratios account for differences in the way control practices affect loadings.

For instance, a pound of phosphorus applied to land far from a waterbody will tend to contribute less to loadings than one applied close to the waterbody. If trading is allowed to occur between these two landowners, an appropriate trading ratio must be established to ensure that the loading goal is met. Specifically, if the landowner near the waterbody is to increase its phosphorus applications, it must purchase reductions from its trading partner at a greater than 1-1 rate. The complexity of watershed interactions does not permit the identification and implementation of such trading ratios. The complexity of watershed systems also contributes to the overall monitoring problem. End-of-pipe monitoring allows for relatively precise monitoring of point source discharges. In general, this kind of precision is not available for nonpoint source loads. The EPA's proposed rules include "margins of safety" to account for the monitoring and causal uncertainties associated with nonpoint sources.

One counter-example is a nonpoint program for phosphorus releases into Lake Okeechobee that relies on nonpoint source monitoring. It is the exception that proves the rule, however. The monitoring takes advantage of the lake system's artificially constructed hydrography, including canals and pumping facilities. Elsewhere, monitoring strategies include proposals for remote-sensing via satellite to determine compliance with land management and construction requirements, such as buffers, cover crops, and irrigation systems. Finally, in some cases, quantitative modelling may be used to indirectly estimate loadings. Results from studies in one area can be used to generalise the relationships between observable practices and typical effects in another. While clearly imperfect, such tools may be the most pragmatic means of injecting some knowledge of control-loading relationships into a trading framework.

While the problems associated with nonpoint monitoring and trading are significant, perhaps even more significant is that water quality trading has to date failed to provide significant benefits even among point sources. The reasons for this are largely legal. Technology-based requirements (and their associated effluent standards) are non-negotiable under the CWA. All point source dischargers must install appropriate treatment mechanisms to achieve these required discharge levels. It is explicit EPA policy that trading participants can in no way be absolved of this baseline technical requirement. While this position may be understandable from an enforcement standpoint, it significantly limits the flexibility over control options. Limited flexibility means that there are limited gains from trading.

Flexibility is also limited in a number of other ways. For instance, point source effluent standards are industry-specific. Since effluent standards are non-negotiable under the CWA, this implies that no trading can occur across firms in different industries.

The practical effect of this lack of flexibility is that a firm seeking a trade may find no other eligible firms with whom to trade. In financial jargon, this is equivalent to a lack of liquidity and undermines the ability of the trading market to reveal and take advantage of cost savings. Another lack of flexibility arises from prohibitions on trading across pollutants. For instance, under the withdrawn new-source offset proposal, the EPA make clear that the offset had to be for the same pollutant.

The informational, scientific, and legal barriers to water quality trading are significant. State experimentation with nonpoint programs and the development of analytical techniques to relate control practices and loadings can be expected. But the history of water quality trading to date urges pessimism regarding the benefits likely to be secured by sophisticated new trading schemes involving point and nonpoint sources. These schemes are largely untried and administratively complex. The search for cost-effective approaches to water quality improvement should not be allowed to rely exclusively on trading. Instead, a variety of more direct forms of regulation, such as mandatory, enforceable BMPs, should also, and perhaps first, be applied to nonpoint sources.

Allocations, Baselines, and Liability

One of the most important policy choices in a trading system is the determination of who is initially responsible for pollutant reductions. Consider two different trading schemes. In the first, point sources are responsible for pollutant reductions but can purchase reductions from other point sources or from nonpoint sources. In the second, both point sources and nonpoint sources are responsible for reductions, but can trade among themselves to achieve those reductions in the most cost-effective manner.

The federal TMDL program leaves the choice of trading program entirely up to the states. If states are to introduce trading, this type of choice will have to be made. At a basic level, this choice is one over the distribution of the costs of discharge reductions. Both systems allow trading, so theoretically both will result in the same least-cost pattern of reductions after trading has occurred. They differ, however, in their allocation of liability for reductions.

In the former case, liability lies with point sources. This means point sources must bear the costs of reducing their own releases or pay nonpoint sources to reduce theirs. In the latter case, liability is shared. Since nonpoint sources are themselves subject to controls they must bear the costs of achieving reductions, or purchase them elsewhere. In addition to determining the initial liability allocation, any trading system must also clearly define baseline (before-the-fact) discharges. This baseline is necessary if trading-driven discharge reductions are to be verified by regulators.

The artificial inflation of baselines is a problem that any trading system must address. Polluters selling credits have an incentive to inflate baseline discharges since inflation increases the amount of discharge reductions they can claim to provide. In the case of point sources, NPDES permits provide a verifiable inventory of baseline releases. Moreover, the point source permitting process counteracts any incentive to overstate releases. Under point source permitting, larger releases tend to imply more stringent, costly control technologies. The nonpoint source situation is quite different. Not having been subject to permit requirements, nonpoint sources lack an independently verifiable, and incentive-compatible baseline. Thus, there will be a problematic tendency for such sources to initially overestimate releases in an attempt to generate larger reduction credits. One way to minimise this kind of problem is to initially allocate responsibility for reductions to nonpoint sources. With nonpoint discharge standards, nonpoint sources have a countervailing incentive to under-claim contributions to loadings. Of course, the larger lesson is that there is the need for independent verification of releases.

The enforcement of a trading system is also directly related to the allocation of liability. For instance, consider the proposed rules' new-source offset requirements. This is a liability allocation akin to the first situation considered above. Point sources are liable for offsetting reductions.

These reductions can be purchased from nonpoint sources. In event of noncompliance by a nonpoint source, the point source is subject to enforcement action, not the nonpoint source. This is true under the EPA's watershed trading framework as well. Consider comments on the EPA's trading framework from one point source which was concerned that "the lack of defined legislation and regulatory controls of nonpoint sources may result in the point source partner of a point source/nonpoint source effluent trading agreement being held solely liable for violations of a water quality standard, even though the source of the violation may be the nonpoint source." However, any point-nonpoint trade presumably involves a contractual agreement between the sources that is independently enforceable under contract law.

While agency enforcement action would be directed at the point source, the point source could exercise its contractual remedy in order to secure the nonpoint source's compliance with contract terms. Nevertheless, point source liability for nonpoint source noncompliance increases the transaction costs associated with trading. This creates yet another argument for direct enforcement of nonpoint controls by states.

The Total Maximum Daily Load program will significantly alter the politics, economics, and implementation of water quality regulation. Improved monitoring of ambient water quality conditions and the accessible public documentation of impairments will focus government and public attention on water conditions that continue to be problematic, even after twenty-five years of CWA regulation. While industrial point sources will no doubt continue to be vivid symbols of the nation's water pollution problems, this image is increasingly inappropriate. Nonpoint agricultural, commercial, and urban sources, while harder to caricature, are the rightful focus of dissatisfaction.

The most powerful aspect of the TMDL rules proposed by EPA is that they are motivated by, and address, water quality issues created by nonpoint sources.

The holistic, watershed-level analysis required by the TMDL process will inevitably identify a larger sphere of often unregulated discharge sources. For these reasons alone, the TMDL program is likely to promote significant, desirable changes in the targets and implementation of water quality regulation. This chapter has provided specific examples of TMDLs and the way in which they are improving the public's knowledge of impairments, motivating new analytic techniques for the identification of sources, and promoting experimentation with new water quality policies. This movement toward a water quality-driven approach marks a welcome, mature phase of water quality regulation.

The changes initiated by the new TMDL rules present a host of challenges. These challenges call for tempered optimism and a willingness to confront the significant implementation issues that will arise from a TMDL based regulatory system. First, there are numerous scientific difficulties associated with creating legally meaningful causal links between dispersed nonpoint sources, their control activities, and changes in surface water pollutant loadings. The causal linkages are poorly understood, and even when developed in the most rigorous manner, only apply to the watershed for which they were specifically developed. The complex and idiosyncratic nature of pollutant discharge, transport, and deposition processes means that the technical underpinnings of TMDLs will be costly.

Administrative costs, together with resistance from currently unregulated sources, will act as a brake on state efforts to propose bold new approaches to control activities. While water quality-based regulation gives point sources an incentive to lobby for nonpoint controls, organised nonpoint interests will undoubtedly continue to resist control requirements. Federal authority to compel, and funding to entice, nonpoint controls is limited. The central role of state law and location-specific political conditions mean that TMDL implementation will be variable across states. Several issues demand greater legal clarity: interjurisdictional conflict is likely, since downstream jurisdictions will often inherit upstream water quality problems; and, air deposition will present particularly knotty jurisdictional issues. The federal role in resolving these inevitable conflicts deserves attention. In addition, the relationship between water quantity allocations and water quality will demand a reconciliation between state quantity and quality laws.

Finally, the rules' emphasis on trading among point sources and nonpoint sources should be viewed as a desirable aspiration, but also as a distraction in the near-term. Significant administrative, monitoring, and enforcement barriers to water quality trading exist. Trading among point sources has to date failed to be practical. Expansion of trading programs to nonpoint sources will only expand the complexity of trading programs. One of the many preconditions for trading is that nonpoint sources be monitorable and that enforcement mechanisms exist to compel corrective actions when nonpoint discharge restrictions are violated. Credibility, transparency, and enforceability are paramount if flexible environmental controls, such as effluent trading, are to be realistically contemplated. Regulatory, legal, and technical efforts should first be directed toward this goal, which is in itself a significant challenge. Despite the challenges it presents, the TMDL approach marks movement toward a welcome, mature phase of water quality regulation. The key feature of EPA's proposed TMDL rules is that they are motivated by, and address, water quality issues created by the widest range of sources. The holistic, watershed-level analysis required by the TMDL process will inevitably identify a larger sphere of often unregulated discharge sources. For these reasons alone, the TMDL program is likely to promote significant, desirable changes in the implementation of water quality regulation.

Bibliography

Alka Rani Upadhyay: *Aquatic Plants for the Waste Water Treatment*, Daya, Delhi, 2004.

Allan, J. A and Chibli Mallat: *Water in the Middle East: legal, political, and commercial implications*, I.B. Tauris, London, 1995.

Anderson, Mary P. and William W. Woessner: *Applied Groundwater Modeling: Simulation of Flow and Advective Transport.* Academic Press, San Diego, 1992.

Asaithambi S. : *Economics of Ground Water Management in India*, Abhijeet, Delhi, 2008.

Asit K. Biswas and Cecilia Tortajada: *Appraising Sustainable Development: Water Management and Environmental Challenges*, Oxford University Press, Delhi, 2005.

Bagis, Ali Ihsan: *Water as an Element of Cooperation and Development in the Middle East*, Ankara, Ayna Publications, 1994.

Bhakar S.R. : *Ground Water Hydrology : Theory and Practice*, Agrotech, Delhi, 2009.

Bhaskara, Harry: *Ending the Daily Water Chore In Against All Odds*, Panos Publications. London. 1989

Bhave, P.R. and R. Gupta: *Analysis of Water Distribution Networks*, Narosa, Delhi, 2011.

Biswas Asit K. : *Integrated Water Resources Management in South and South-East Asia*, , Oxford University Press, Delhi, 2001.

Bourne, Peter G.: *Water and Sanitation: Economic and Sociological Perspectives* Academic Press. Orlando. 1984.

Boyer , J.S.: *Measuring the Water Status of Plants and Soils*, Academic Press, N.Y., 1995.

Brooks, D.: *Water: Local-Level Management*, Ottawa, International Development Research Centre, 2002.

Bullock, John and Darwish, Adel: *Water Wars: Coming Conflicts in the Middle East*, London, Victor Gollancz, 1999.

Cazmuri, Renato : *"Chilean Water Policy Experience"*, Agriculture and Natural Resources Department, World Bank, Washington, D.C. Processed, 1992.

Conway, D.: *Climate Change and Water Resources in the Nile Basin*, University of London, London, 1993.

Cullis, A. : *Rainwater Harvesting: The Collection of Rainfall and Runoff in Rural Areas,* London, U.K.: IT Publications, 1986.

D'Itri, Frank M. and Lois G. Wolfson: *Rural Groundwater Contamination.* Lewis Publishers, Chelsea, 1987.

Doneen L.D.: *Water Quality for Agriculture,* Department of Irrigation, University of California, Davis, 1964.

Elmusa, Sharif S.: *Water Conflict: Economics, Politics, Law and Palestinian-Israeli Water Resources,* Institute for Palestine Studies, Beirut, Lebanon, 1998.

Engelman, R., and P. LeRoy : *Sustaining Water Washington,* D.C.: Population Services International, New York, 1993.

Erik Nissen-Peterson : *Rainwater Catchment Systems,* UK: Intermediate Technology Publications, 1999.

Feder, Paul A.: *A Guidebook to Groundwater: Resources and Education Opportunities in the Great Lakes Region,* Great Lakes Commission, Ann Arbor, 1993.

Freeze, R. Allan and John A. Cherry: *Groundwater.* Prentice Hall, Englewood Cliffs, NJ. 1979.

Garfinkle, Adam M.: *War, Water, and Negotiation in the Middle East: the Case of the Palestine-Syria Border 1916-1923,* Tel Aviv University, Tel Aviv, 1994.

Ghosh, N.C. and K.D. Sharma: *Groundwater Modelling and Management,* Capital Pub, Delhi, 2006.

Gould, J., and E. Nissen-Petersen : *Rainwater Catchment Systems for Domestic Supply: Design, Construction and Implementation,* London, U.K.: IT Publications, 1999.

Gupta K.R.: *Water Crisis in India,* Atlantic Pub, Delhi, 2008.

Gupta, O.P. : *Water in Relation to Soils and Plants : With Special Reference to Agriculture,* Agrobios, Delhi, 2002.

Heath, Ralph C.: *Basic Ground-water Hydrology,* Denver, CO: U.S. Geological Survey, 1983.

Helen Bannayan: *Water Resources of Jordan: Present Status and Future Potentials,* Amman, Friedrich-Ebert-Stiftung, 1993.

Holubeshen, Moufida; Amery, Hussein A.: *Annotated Bibliography of Water in the Middle East,* University of Toronto, Toronto, 1995.

Husain, Ahmad : *Environment and Water Resource Management,* Sumit Enterprises, Delhi, 2006.

Jana B L : *Water Harvesting and Watershed Management,* Agrotech, Delhi, 2008.

Jat, M.L. and S.R. Bhakar: *Ground Water Hydrology : Theory and Practice,* Agrotech, Delhi, 2009.

Jee Chandrawati and Shagufta : *Rainwater Harvesting*, A.P.H. Pub, Delhi, 2010.

Kally, Elisha and Gideon Fishelson: *Water and Peace: Water Resources and the Arab-Israeli Peace Process*, Westport, CT, Praeger, 1993.

Kanmony, J. Cyril : *Drinking Water Management : Problems and Prospects*, Mittal Pub, Delhi, 2010.

Khan M A : *Water Resources Management and Sustainable Agriculture*, APH, Delhi, 2008.

LeRoy, K. : *Sustaining Water Washington*, D.C.: Population Services International, New York, 1993.

Libiszewski, Stephan: *Water Disputes in the Jordan Basin Region and their Role in the Resolution of the Arab Israeli Conflict*, Center for Security Studies and Conflict Research, Zurich, 1995.

Linda M. Welkom: *Groundwater Chemicals Desk Reference*, Chelsea, MI, 1990.

Lloyd Myers : *Handbook of Water Harvesting*, Washington D.C.: U.S. Dept. of Agriculture, Agricultural Research Service, 1983

Lowi, Miriam: *Water and Power: the Politics of a Scarce Resource in the Jordan River Basin*, Cambridge University Press, Cambridge, Eng, 1993.

Mahajan Gautam : *Evaluation and Development of Ground Water*, APH, Delhi, 2008.

Matthess, Georg : *The Properties of Groundwater*, John Wiley & Sons, New York, NY. 1982.

McGarry, M.G.: *Matching Water Supply Technology to the Needs and Resources of Developing Countries,* United Nations. New York. 1987

Meenu Bhatnagar: *Groundwater Management : Sustainable Approaches*, Icfai Books, Delhi, 2012.

Metcalf and Eddy Inc: *Waste-water Engineering: Treatment, Disposal, Reuse*, McGraw-Hill Book Co, New York, 1979.

Mishra Archana : *Water Harvesting : Ecological and Economic Appraisal*, Authors Press, Delhi, 2006.

Montgomery, John H. and Linda M. Welkom: *Groundwater Chemicals Desk Reference*, Chelsea, MI, 1990.

Murakami, Masahiro, *Managing Water for Peace in the Middle East : Alternative Strategies*, Tokyo and NYC, United Nations University Press, 1995.

Naff T. and Matsow R.: *Water in the Middle East: Conflict or Cooperation*, Boulder, 1990.

Nielsen, David M.: *Practical Handbook of Groundwater Monitoring*. Lewis Publishers, Chelsea, 1991.

Nissen-Petersen, E. : *Rainwater Catchment Systems for Domestic Supply: Design, Construction and Implementation,* London, U.K.: IT Publications, 1999.

Pacey A., and A. Cullis : *Rainwater Harvesting: The Collection of Rainfall and Runoff in Rural Areas,* London, U.K.: IT Publications, 1986.

Pahwa Prem S. *: Water Harvesting, Purification and Distribution Management,* Dominant, Delhi, 2001.

Palanisami, K : *Groundwater Management and Policies,* Macmillan Publishers India, Delhi, 2008.

Parikshit Ballabh: *Water Resource Management : Planning and Development,* Cyber Tech, Delhi, 2009.

Premjit Sharma: *Agricultural Drainage and Water Quality,* Gene Tech Books, Delhi, 2007.

Ranghaswami, M.V. ; K. Palanisami: *Groundwater Resources Assessment, Recharge and Modelling,* Macmillan Publishers India, Delhi, 2008.

Rao K. Nageswara *: Water Resources Management : Realities and Challenges,* New Century Publication, Delhi, 2006.

Rowe, Garry W. and Sylvia J. Dulaney: *Building and using a Groundwater Database.* Lewis Publishers, Chelsea, 1991.

Salameh, Elias and Helen Bannayan: *Water Resources of Jordan: Present Status and Future Potentials,* Amman, Friedrich-Ebert-Stiftung, 1993.

Sanghvi Sheela and Pahwa Prem S. *: Water Harvesting, Purification and Distribution Management* , Dominant, Delhi, 2001.

Sharma, R. K. and T.K.Sharma: *A Textbook Of Water Power Engineering,* S. Chand Publisher, Delhi, 2008.

Sharma, S.S.P. and U.H. Kumar: *Dynamics of Watershed Development and Livelihood in India,* Serials Pub, Delhi, 2010.

Singh A K *: Environment and Water Resources Management,* Adhyayan, Delhi, 2006.

Thapliyal, B K : *Democratisation of Water,* Serials Pub, Delhi, 2008.

Thomas, Harold E.: *The Conservation of Groundwater: A Survey of the Present Groundwater Situation in the U.S.,* McGraw Hill, New York, NY. 1951.

Tideman, E. M.: *Watershed Management – Guidelines for Indian Conditions,* Omega Scientific Publishers, Delhi, 1996.

Todd, David Keith: *Groundwater Hydrology.* John Wiley & Sons, New York, NY. 1959.

Van der Leeden, Frits: *Ground Water: A Selected Bibliography,* Port Washington, NY, 1974.

Walton, William C.: *Groundwater Resource Evaluation.* McGraw Hill, New York, 1970.

Index

N

O

P

R